BASIC
FUNDAMENTALS OF MATH

for

ADDITION, SUBTRACTION, MULTIPLICATION AND DIVISION

using

WHOLE NUMBERS, FRACTIONS, DECIMALS AND PERCENTS

by

Jerry Ortner

AuthorHouse™
1663 Liberty Drive
Bloomington, IN 47403
www.authorhouse.com
Phone: 1-800-839-8640

First published by AuthorHouse 7/26/2011

ISBN: 978-1-4634-4236-1 (sc)
ISBN: 978-1-4634-4235-4 (e)

Library of Congress Control Number: 2011912752

Printed in the United States of America

*Any people depicted in stock imagery provided by Thinkstock are models,
and such images are being used for illustrative purposes only.
Certain stock imagery © Thinkstock.*

This book is printed on acid-free paper.

TABLE OF CONTENTS

LESSON 1 – Place Values

Trillions		Hundred-billions	Ten-billions	Billions		Hundred-millions	Ten-millions	Millions		Hundred-thousands	Ten-thousands	Thousands		Hundreds	Tens	Ones
__	,	__	__	__	,	__	__	__	,	__	__	__	,	__	__	__

Practice on place values.

1. Find the place value of the digit 3 in the number 5,238,876,914,684
   - a. billions
   - b. ten-billions
   - c. ten- millions
   - d. tens

2. Find the place value of the digit 7 in the number 9,238,861,754,184
   - a. ten-thousand
   - b. million
   - c. ten-billion
   - d. hundred-thousand

3. Find the place value of the digit 2 in the number 8,374,465,129,549
   - a. trillion
   - b. hundred
   - c. million
   - d. ten-thousand

4. Find the place value of the digit 7 in the number 5,642,219,873,923
   - a. hundred-billion
   - b. hundred-million
   - c. ten-thousand
   - d. million

5. What is the place value of the 7 in 9,235,546,718,658?

6. What is the place value of the 9 in 5,386,694,217,467?

7. What is the place value of the 6 in 8,967,721,435,175?

8. What is the place value of the 7 in 6,354,421,978,148?

9. Find the place value of the digit 3 in the number 6,158,849,327,987
   - a. hundred-thousand
   - b. trillion
   - c. million
   - d. hundred

10. Find the place value of the digit 9 in the number 8,654,497,123,743
    - a. ten-thousand
    - b. million
    - c. ten-billion
    - d. ten-million

LESSON 2 – Rounding

Look at the question. Decide on what place value the number is to be rounded to and look **_one place to the right_** of that. If less than 5, do not round up and place zeros for each location in the number. If 5 or more, increase the place value by 1 and add zeros for fillers.

Practice on rounding.

1. Round 9138 to the nearest thousands.
                    a. 1000   b. 9100   c. 9000   d. 9140

2. Round 2663 to the nearest thousands.
                    a. 3000   b. 1000   c. 2700   d. 2660

3. Round 5215 to the nearest thousands.
                    a. 1000   b. 5000   c. 5220   d. 5200

4. Round 45,169 to the nearest tens.

5. Round 58,276 to the nearest hundreds.

6. Round 46,482 to the nearest thousands.

7. Round 1634 to the nearest tens.
                    a. 1000   b. 1630   c. 1634   d. 1600

8. Round 4547 to the nearest hundreds.
                    a. 4500   b. 5000   c. 4550   d. 4600

9. Round 7758 to the nearest thousands.
                    a. 8000   b. 7760   c. 7000   d. 7800

10. Round 57,373 to the nearest thousands.

11. Round 47,266 to the nearest hundreds.

12. Round 62,187 to the nearest ten-thousands.

13. Round 4727 to the nearest tens

14. Round 9341 to the nearest hundreds.

15. Round 1823 to the nearest tens.
    a. 1830   b. 2000   c. 1800   d. 1820

16. Round 2752 to the nearest thousands.
    a. 1000   b. 2800   c. 3000   d. 2750

17. Round 8141 to the nearest hundreds.
    a. 8200   b. 8100   c. 8140   d. 8000

18. Round 68,289 to the nearest thousands.

19. Round 44,466 to the nearest hundreds.

20. Round 6,499,990 to the nearest millions.

21. Round 74,195 to the nearest hundreds.

22. Round 23,681 to the nearest ten-thousands.
    a. 23,700   b. 30,000   c. 20,000   d. 24,000

23. Round 2175 to the nearest tens.
    a. 2170   b. 2180   c. 2200   d. 1000

24. Round 4444 to the nearest thousands.
    a. 4440   b. 5000   c. 4000   d. 4400

25. Round 64,372 to the nearest hundreds.

26. Round 86,464 to the nearest thousands.

27. Round 146,501 to the nearest thousands.

28. Round 7,500,000 to the nearest millions.

29. Round 3456 to the nearest tens.

30. Round 15,675 to the nearest hundreds.

31. Round 43,875,432 to the nearest ten-thousands.

32. Round 23,484 to the nearest hundreds.

33. Round 175,458 to the nearest thousands.

LESSON 3 – Subtraction of Whole Numbers

There is no problem when the number on the top (the minuend) is equal to or larger than the number on the bottom (the subtrahend).

Examples:

$$\begin{array}{r} 56 \\ -\ 23 \\ \hline 33 \end{array} \qquad \begin{array}{r} 147 \\ -\ 26 \\ \hline 121 \end{array} \qquad \begin{array}{r} 596 \\ -\ 345 \\ \hline 251 \end{array} \qquad \begin{array}{r} 4896 \\ -\ 2372 \\ \hline 2524 \end{array}$$

However, everything changes when the bottom number is greater than the top.

Example:

$$\begin{array}{r} 52 \\ -\ 27 \\ \hline \end{array}$$

In this problem, you cannot subtract 7 from 2. So you need to borrow a "10" from the 50 and add that 10 to the 2, which now becomes 12 and the 5 reduces to a 4.

$$\begin{array}{cc} 4 & 12 \\ \cancel{5} & \cancel{2} \\ -2 & 7 \\ \hline 2 & 5 \end{array}$$

Now the top number is larger than the bottom, so we may now proceed.

Check your work by adding the answer (the difference) and the minuend.
25 + 27 = 52 ✓ AOK!

Let's try another.

$$\begin{array}{ccc} 3 & 2 & 4 \\ -1 & 5 & 8 \\ \hline \end{array}$$

First borrow 10 from the 2 so that the 4 becomes 14, and the 2 becomes a 1.

$$\begin{array}{ccc} & 1 & 14 \\ 3 & \cancel{2} & \cancel{4} \\ -1 & 5 & 8 \\ \hline & & 6 \end{array}$$

Now subtract 8 from 14. The result is 6. Moving over one position to the left, we can't subtract 5 from 1. So we have to borrow 10 from the left.

$$\begin{array}{ccc} 2 & 11 & 14 \\ \cancel{3} & \cancel{2} & \cancel{4} \\ -1 & 5 & 8 \\ \hline 1 & 6 & 6 \end{array}$$

Then 11 – 5 = 6 and finally 2 – 1 = 1.

Check?? Does 166 + 158 = 324?

Try another with zeros on top!

$$\begin{array}{r} 4003 \\ -\ 1462 \\ \hline \end{array}$$

First subtract 2 from 3. It can be done.
Result = 1.

4

```
    4   0   0   3
  − 1   4   6   2
  _____
                    1
```

Now we can't subtract 6 from zero (0), so we need to borrow. The next number to the left is another 0. We need to go over one more place to the left. Reduce the 4 to a 3. Place a 10 in the hundreds place. Still 0 − 6. So we need to borrow again.

```
  3   10
  4    0   0   3
− 1    4   6   2
  _____
  2    5   4   1
```

The 10 now becomes a 9 and the 0 becomes a 10. Now we have 10 − 4.

```
          9
  3      10   10
  4       0    0   3
− 1       4    6   2      9 − 4 = 5   and   3 − 1 = 2
  _____
  2       5    4   1
```

Check to make sure the answer is correct:
2541 + 1462 = 4003 ✓

Try another with words rather than numbers.

What number is twelve less than forty?

Note: Less than indicated the subtraction operation.

Rewrite the statement with numbers: 40 − 12 = ☐?

```
  3    10    Can't subtract 2 from zero (0), so we need to borrow 10
  4     0    from 4. It becomes a three and the zero becomes 10.
− 1     2
  _____
  2     8    10 − 2 = 8  and  3 − 1 = 2

            Check!! 28 + 12 = 40 ✓
```

Now try these practice problems to see if you can master subtraction.

1.   4000 – 1279   a. 2731   b. 2711   c. 2821   d. 2721

2.   3000 – 1507   a. 1393   b. 1493   c. 1503   d. 1593

3.   What number is twenty-nine less than forty-four?

4.   What number is eighteen less than fifty?

5.   7835 – 243

6.   Subtract: 7000 – 1858

7.   Subtract: 4000 – 1030

8. $6.15
       – $0.98   a. $5.02   b. $6.17   c. $7.13   d. $5.17

9. $5.22
       – $0.98   a. $5.24   b. $6.66   c. $4.24   d. $5.09

10. Subtract: 8000 – 4333

11. 6526 – 294

12. 5876 – 392

13. What number is thirty-one less than sixty-six?

14. 4000 – 2757   a. 1143   b. 1253   c. 1343   d. 1243

15. 3000 – 1279   a. 1821   b. 1721   c. 1731   d. 1711

16. 5203 – 2847

17. Subtract 541 from 8470.

18. 10,500 – 4587

19. What number is 42 less than 100?

20. 234,567 – 123,988

21. Fifteen is how much less than 90?

22. $150.00 - $99.95

LESSON 4 – Multiply and Divide Numbers
Use of calculator is permitted.

1. What is the product of three thousand seven hundred forty-seven and twenty-one?

2. What is the product of two thousand seven hundred twenty-one and seventeen?

3. 126 × 478   a. 59,228   b. 60,128   c. 60,218   d. 60,228

4. 136 × 337   a. 45,822   b. 45,832   c. 44,832   d. 45,732

5. 147 × 209   a. 30,713   b. 30,723   c. 30,623   d. 29,723

6. What is the product of one thousand nine hundred sixty-three and thirty-four?

7. Multiply: 35 × 41

8. Rewrite the addition problem as a multiplication problem and solve.   (Different way to express addition.)

$$382 + 382 + 382 + 382 + 382 + 382$$

9.   835
  × 49

10. Multiply: $3.63
     × 60   a. $217.80 b. $435.60 c. $207.80 d. $63.63

11. 23 $\overline{)769}$ a. 32 R 6   b. 33 R 10   c. 32 R 11   d. 33 R 14

12. 33 $\overline{)536}$ a. 16 R 22   b. 16 R 8   c. 15 R 18   d. 15 R 31

13. Divide: 3596 ÷ 49

14. Divide: 5476 ÷ 64

15. Divide: $\dfrac{648}{81}$

16. If 132 students are divided into groups of 11, how many groups will be formed?

17. Divide: $\dfrac{957}{33}$

18. Divide: $\dfrac{504}{56}$

19. 3795 ÷ 87

20. 1855 ÷ 72

21. If 150 players are separated into 15 equal teams, how many players will there be on each team?

22. 6$\overline{)6291}$

23. Here are three ways to write 72 divided by 8 (Three ways to show division problems.):

$$8\overline{)72} \qquad 72 \div 8 \qquad \dfrac{72}{8}$$

Show three ways to write 54 divided by 6.

24. A discount audio store advertised a collection of country music on 11 compact discs for $56.65. What is the cost of each disc?
    a. $5.15   b. $5.26   c. $5.17   d. $5.25

25. If 315 players are separated into 21 equal teams, how many players will there be on each team?

LESSON 5 – Word Problems Focusing on Addition and Subtraction

1. Karl bought a pair of slacks for $33.25 and a shirt for $13.73. What was the total cost of these items?

    a. $46.98   b. $47.98   c. $47.88   d. $47.48

2. Cayla bought a skirt for $41.12 and a blouse for $12.52. What was the total cost of these items?

    a. $52.64   b. $53.64   c. $54.14   d. $52.54

3. What number is twenty-seven less than seventy-nine?

4. Rose paid $10 for a $9.18 book. How much money should she get back?

5. Fred paid $5 for a $1.42 lunch. How much money should he get back?

6. The band club sold 3119 candy bars the first week of the fundraiser to buy new band uniforms. 3365 candy bars were sold the second week. How many candy bars were sold the first two weeks?

7. A soccer stadium holds 24,000 people. If 7000 people can be seated in the bleachers, how many seats are available in the rest of the stadium?

8. The chorus sold 1433 candy bars the first week of the fundraiser. 5994 candy bars were sold the second week. How many candy bars were sold the first two weeks?

    a. 7427 candy bars   b. 7027 candy bars
    c. 7407 candy bars   d. 7424 candy bars

9. A football stadium holds 17,000 people. If 5000 people can be seated in the bleachers, how many seats are available in the rest of the stadium?

10. Carl bought a pair of sneakers for $64.00 and socks for $12.50. What was the total cost of these items?

    a. $51.50   b. $64.00   c. $76.50   d. $12.50

11. A six-pack of soda costs $2.49 and a dozen donuts are $5.50. How much change do you receive if you buy these items and hand a ten-dollar bill to the clerk?

12. What is the least number of coins for receiving change when the item is 59¢ and you give the clerk a $1 bill?

13. Three six-packs of soda cost $12.00. How much change do you receive giving the clerk a $20 bill?

14. Twenty-three guys are selected to play in a scrimmage. To insure you get more players than the other team, do you select first or second each round?

LESSON 6 – Add, Subtract, and Multiply Decimals

When adding decimals, keep the decimal points in a straight line (vertically). You can fill in zeros if you wish. Then add each column, starting at the right and moving left. Likewise, when subtracting, decimal beneath decimal.

Example 1:   2.3 + 4.363 + 5.12 + 7 + 8.132

```
    2.300
    4.636
    5.120
    7.000
 +  8.132
   27.182
```

Example 2:   24.02 – 15.147

```
        13      9  11
     1   3     10  1  10
     2   4  .  0   2  0
  -  1   5  .  1   4  7
     8   .  8   7  3
```

↑

OH!! Don't forget the decimal point in your answer!

When multiplying, count off the places in the answer.

Example 3:   16.24 × 1.3

```
      1  6  .  2  4  ←  2 decimal places   total of 3 decimal
   ×        1  .  3  ←  1 decimal place    places
      4  8  7  2
   1  6  2  4
   2  1 . 1  1  2  ←  starting from the far right, move the
                      decimal point 3 places to the left.
```

Therefore, 16.24 × 1.3 = 21.112.

Practice problems on adding, subtracting, and multiplying decimals.

1.   2.8 + 0.3 + 2.3   a. 12.4   b. 10.4   c. 4.9   d. 5.4

2.   0.6 + 0.32 + 5.32   a. 5.94   b. 6.24   c. 10.24   d. 15.24

3.   Add: 1.84 + 2.22 + 4.71

4.   Subtract: 39.4 – 7.32

5.   34.3 – 5.95

6.   9.81 + 8 + 8.1   a. 25.92   b. none of these   c. 26.91   d. 26.01

7.   30.7 – 18   a. 28.9   b. 1.27   c. 12.7   d. 289.0

8.   2.48 + 8 + 4.9

9.   What is the sum of 2.18, 0.465, and 6? (sum means to add)

10. 0.76 × 0.01

11. 0.56 × 0.06

12. 0.89 × 0.07   a. 0.00623 b. 6.23   c. 0.623   d. 0.0623

13. 0.5 × 0.6 × 0.3

14. 0.6 × 0.2 × 0.7

15. 0.86 × 0.06

16. 0.88 × 0.09

17. 7 – (2.29 + 0.61)   add the two numbers inside the parentheses before subtracting

18. 16.23 + 8.4 + 702.31

19. 1.23 × 40.4

20. 236.9 – 178.37

21. (4.64 + 7.27) – (3.14 + 6.4)

LESSON 7 – Dividing Decimals Using Whole Numbers as Divisors

Some examples to start.

Example 1:   3.84 ÷ 4

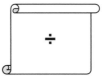

If the divisor is a whole number, place the decimal directly above its location in the dividend.

```
        .  9  6
          ↑
    4 | 3 .  8  4
        3     6
        ─────────
              2  4
              2  4
              ─────
                 0
```

Example 2:   0.0245 ÷ 7

Make certain the placeholders (zeros) are inserted to the right of the decimal point when you can't divide.

```
      0 .  0  0  3  5
    7 | 0 .  0  2  4  5
                2  1
              ─────────
                 3  5
                 3  5
                 ─────
                    0
```

Example 3:   256.44 ÷ 24

Add zeros to the dividend so the problem can be completed.

```
          1  0 .  6  8  5
    24 | 2  5  6 .  4  4  0
         2  4
         ─────
            1  6     4
            1  4     4
            ──────────
               2     0  4
               1     9  2
               ──────────
                     1  2  0
                     1  2  0
                     ────────
                           0
```

Practice problems on dividing decimals.

1. $0.84 \div 4$

2. $9 \overline{)6.3}$

3. $6 \overline{)0.36}$

4. $2.46 \div 6$

5. $2.13 \div 3$

6. $4 \overline{)0.24}$ a. 0.06   b. 0.9   c. 0.09   d. 0.6

7. $5 \overline{)0.35}$ a. 0.7   b. 0.9   c. 0.09   d. 0.07

8. $1.68 \div 4$

9. $4.97 \div 7$

10. $1.26 \div 6$

11. $2.05 \div 5$

12. $1.28 \div 4$

13. $3 \overline{)0.12}$ a. 0.4   b. 0.04   c. 0.06   d. 0.6

14. $7 \overline{)2.1}$ a. 0.03   b. 0.02   c. 0.2   d. 0.3

15. $7 \overline{)0.35}$ a. 0.5   b. 0.7   c. 0.05   d. 0.07

16. $5 \overline{)0.3}$ a. 0.06   b. 0.7   c. 0.6   d. 0.07

17. $4 \div 2000$

18. $6 \overline{)0.36}$

19. $9.2 \div 4$

20. $8\overline{)0.736}$

21. $32\overline{)12}$

22. $91.44 \div 36$

23. $2\overline{)5.328}$

24. $121.44 \div 24$

25. $48\overline{)60}$

26. $8\overline{)10,009.6}$

27. $1904.76 \div 26$

28. $9\overline{)7.2}$ a. 0.08   b. 0.8   c. 8   d. 0.008

29. $16 \div 25$

30. $30\overline{)849.3}$

31. $1140.48 \div 8$

32. $9\overline{)573.48}$

33. $3000 \div 4$

34. $4 \div 2000$

35. $336.8592 \div 6$

LESSON 8 – Converting Decimals to Fractions and Vice Versa

Here again place value comes in handy. Go to Lesson 1 to review place values. To the right of the decimal point, the place values are as follows.

.　tenths　hundredths　thousandths　ten-thousandths

Several examples converting a decimal to a common fraction:

Example 1:　　0.3 = $\frac{3}{10}$

Example 2:　　0.46 = $\frac{46}{100}$　　(which reduces to $\frac{23}{50}$)

Example 3:　　0.571 = $\frac{571}{1000}$

A good way to check is to look at the denominator of the fraction. It should have as many zeros as places to the right of the decimal point. In example 1, there is one place to the right of the decimal point so there will be one zero in the denominator of the fraction. Example 2, two places so two zeros, etc.

Also remember to be able to work with a whole number and decimal.

Example 4:　　23.67 = $23\frac{67}{100}$

Example 5:　　1234.56789 = $1234\frac{56789}{100,000}$

Converting a fraction to a decimal is the reverse operation. Count the number of zeros in the denominator. That is, how many zeros there are after the number 1. (Note: make certain that the denominator is in tenths, hundredths, thousandths, etc)

Example 6:　　What is $\frac{72}{100}$ as a decimal? Answer: 0.72

Example 7:　　What is $1\frac{37}{1000}$ as a decimal? Answer: 1.037

Example 8: What is $\frac{9}{10}$ as a decimal? Answer: 0.9

Example 9: What is $\frac{125}{500}$ as a decimal? Answer: 0.25

Convert: $\frac{125}{500} = \frac{x}{1000}$ cross-multiply to solve: x = 250

(Note: the zero after the 5 is not necessary in the decimal answer.)

Now try the practice problems and see if you understand this lesson thoroughly.

1. Write 0.59 as a common fraction.
    a. $\frac{590}{100}$ b. $\frac{5.9}{100}$ c. $\frac{59}{100}$ d. $\frac{59}{10}$

2. Write 0.45 as a common fraction.
    a. $\frac{4.5}{100}$ b. $\frac{45}{100}$ c. $\frac{45}{10}$ d. $\frac{450}{100}$

3. Write 0.83 as a common fraction.

4. Write 0.27 as a common fraction.

5. Write $\frac{50}{100}$ as a decimal number.

6. Write 0.11 as a common fraction.

7. Write 0.47 as a common fraction.

8. Write 0.65 as a common fraction.
    a. $\frac{6.5}{100}$ b. $\frac{650}{100}$ c. $\frac{65}{10}$ d. $\frac{65}{100}$

9. What is $\frac{33}{1000}$ as a decimal?
    a. 0.0033   b. 0.033   c. 3.3   d. 0.33

10. What is $\frac{21}{10}$ as a decimal?
    a. 2.1   b. 0.0021   c. 0.21   d. 0.021

11. What is $\frac{27}{1000}$ as a decimal?
    a. 2.7   b. 0.0027   c. 0.27   d. 0.027

12. What is $\frac{3}{10}$ as a decimal?
    a. 0.003   b. 0.03   c. 0.3   d. 0.0003

13. Write 0.39 as a common fraction.

    a. $\frac{3.9}{100}$ b. $\frac{390}{100}$ c. $\frac{39}{10}$ d. $\frac{39}{100}$

14. Write 0.33 as a common fraction.

    a. $\frac{3.3}{100}$ b. $\frac{33}{100}$ c. $\frac{330}{100}$ d. $\frac{33}{10}$

15. Write 0.79 as a common fraction.

16. Write 0.29 as a common fraction.

17. Write 0.75 as a common fraction.

    a. $\frac{75}{100}$ b. $\frac{7.5}{100}$ c. $\frac{75}{10}$ d. $\frac{750}{100}$

18. Write 0.25 as a common fraction.

    a. $\frac{250}{100}$ b. $\frac{2.5}{100}$ c. $\frac{25}{10}$ d. $\frac{25}{100}$

19. Write 0.43 as a common fraction.

    a. $\frac{4.3}{100}$ b. $\frac{43}{10}$ c. $\frac{43}{100}$ d. $\frac{430}{100}$

20. What is $\frac{16}{1000}$ as a decimal?

    a. 0.16   b. 0.016   c. 0.0016   d. 1.6

21. What is $\frac{4}{10}$ as a decimal?

    a. 0.004   b. 0.0004   c. 0.04   d. 0.4

22. Write 0.77 as a common fraction.

    a. $\frac{77}{100}$ b. $\frac{77}{10}$ c. $\frac{7.7}{100}$ d. $\frac{770}{100}$

23. Write 0.55 as a common fraction.

    a. $\frac{55}{10}$ b. $\frac{5.5}{100}$ c. $\frac{55}{100}$ d. $\frac{550}{100}$

24. Write 0.73 as a common fraction.

25. Write 0.37 as a common fraction.

26. Write $\frac{70}{100}$ as a decimal number.

27. Write 43.65 as a common fraction.

28. Write $16\frac{2}{5}$ as a decimal.

29. Write 123.45 as a common fraction.

LESSON 9 – Divisibility Tests

The Rules for Divisibility are:

divisible by 2	even number
divisible by 3	the sum of the digits is divisible by 3
divisible by 4	last 2 digits are a multiple of 4, divide last two digits by 4, if no remainder, then divisible by 4
divisible by 5	last digit is a 0 or 5
divisible by 6	the sum of the digits is divisible by 3 AND an even number
divisible by 8	the last three digits of the number is divisible by 8
divisible by 9	the sum of the digits is divisible by 9
divisible by 10	last digit is a 0

In certain textbooks, divisibility is shown as 3|1568. Is 1568 divisible by 3? Another example: 6|8142. Is 8142 divisible by 6? The answer to these are: 3|1568 is no and 6|8142 is yes.

Some practice problems to try.

1. Is 426 divisible by both 5 and 10?

2. Which of these numbers is divisible by both 9 and 2?
      a. 126   b. 567   c. 117   d. 22

3. Which of these numbers is divisible by both 10 and 3?
      a. 1301   b. 71   c. 390   d. 52

4. Which of these numbers is divisible by both 3 and 9?
      a. 40   b. 62   c. 189   d. 98

5. Which of these numbers is divisible by both 9 and 10?
      a. 64   b. 1170   c. 1054   d. 171

6. Which of these numbers is divisible by both 10 and 5?
      a. 1701   b. 111   c. 176   d. 850

7. Which of these numbers is divisible by both 3 and 2?
      a. 66   b. 51   c. 99   d. 52

8. Is 116 divisible by both 3 and 9?

9. Is 390 divisible by both 10 and 3?

10. Is 566 divisible by both 9 and 10?

11. Is 32 divisible by both 3 and 10?

12. Is 195 divisible by both 5 and 3?

13. Which of these numbers is divisible by both 3 and 5?
                                 a. 153   b. 33   c. 175   d. 255

14. Which of these numbers is divisible by both 9 and 3?
                                 a. 566   b. 189   c. 34   d. 118

15. Which of these number is divisible by both 10 and 2?
                                 a. 220   b. 27   c. 171   d. 1101

16. Which of these numbers is divisible by both 2 and 10?
                                 a. 15   b. 53   c. 260   d. 171

17. Is 117 divisible by both 3 and 2?

18. Is 129 divisible by both 5 and 10?

19. Is 109 divisible by both 10 and 9?

20. Is 70 divisible by both 2 and 5?

21. 4|15,984

22. 5|38,814

23. 6|104,538

24. 8|28,096

25. 9|11,378

26. 3|5958

27. 4|10,612

28. 5|48,670

29. 9|23,772

30. 6|163,944

LESSON 10 – Prime and Composite Numbers

By definition, a prime number has only itself and one as its factors. All others are composite. The prime numbers less than 20 are 2, 3, 5, 7, 11, 13, 17, and 19.

Practice problems on prime and composite numbers.

1.   Which of the numbers is prime: 21, 57, 39, 31

2.   Which of the numbers is prime: 15, 35, 83, 25

3.   Which of the following is a prime number?
                     a. 27   b. 25   c. 67   d. 33

4.   Which of the following is a prime number?
                     a. 21   b. 73   c. 39   d. 15

5.   Which of the numbers is prime: 57, 73, 25, 27

6.   Which of the numbers is prime: 35, 51, 47, 21

7.   Which of the following is a prime number?
                     a. 27   b. 89   c. 9   d. 57

8.   Which of the following is a prime number?
                     a. 29   b. 33   c. 35   d. 21

9.   Write the composite numbers from the list: 19, 29, 18, 37, 21, 26, 43, 47, 10

10.   Write the composite numbers from the list: 5, 12, 13, 16, 19, 18, 29, 6, 37

11.   How many of the following numbers are composite: 12, 36, 14, 4, 32, 55
                     a. 5   b. none   c. 4   d. 6

12. How many of the following numbers are composite: 28, 43, 24, 22, 38, 18, 8, 54
            a. 5   b. none   c. 8   d. 7

13. How many of the following numbers are composite: 42, 51, 3, 26, 15, 52, 19, 33
            a. 8   b. 6   c. 5   d. 3

14. Write the composite numbers from the list: 29, 37, 43, 22, 26, 47, 2, 32, 12

15. How many of the following numbers are composite: 31, 22, 20, 24, 15, 41, 3, 28
            a. 5   b. none   c. 8   d. 4

16. List the first 10 composite numbers.

17. List all composite numbers between 20 and 30.

18. List all prime numbers between 20 and 50.

19. Which of these numbers are composite: 8, 35, 47, 77, 67, 54, 33, 71, 19, 57

20. What are the composite numbers between 60 and 70?

21. How many prime numbers are there between 12 and 25? What are they?

22. True or false: When you add two prime numbers, the sum is always an even number.

23. How many single-digit composite numbers are there? What are they?

24. List the single-digit numbers.

25. Name the first two composite numbers that are **NOT** prime.

26. When you multiply any two prime numbers, is the result (product) always prime or composite?

LESSON 11 – Prime Factorization of a Number

Having found the first ten prime numbers, we can move forward with factorization of any number.

Let's take this example: 216. This is how I figure out prime factors of a number:

$$2^3 \begin{cases} 2 & | & 216 \\ 2 & | & 108 \\ 2 & | & 54 \end{cases}$$
$$3^3 \begin{cases} 3 & | & 27 \\ 3 & | & 9 \\ 3 & | & 3 \end{cases}$$
$$1$$

We finally get $2^3 \cdot 3^3$. Both 2 and 3 are prime numbers. Therefore the prime factorization of 216 is $2^3 \cdot 3^3$.

Try another with me before you tackle the problems below. The number is 180.

$$2^2 \begin{cases} 2 & | & 180 \\ 2 & | & 90 \end{cases}$$
$$3^2 \begin{cases} 3 & | & 45 \\ 3 & | & 15 \end{cases}$$
$$5 \begin{cases} 5 & | & 5 \end{cases}$$
$$1$$

The prime factorization of 180 is $2^2 \cdot 3^2 \cdot 5$.

Try these practice problems. If the answer is given, multiply the prime numbers to ascertain its product. Good luck!

1. Determine the prime factorization of 2100.
    - a. $2 \cdot 2 \cdot 3 \cdot 5 \cdot 5 \cdot 7$
    - b. $2 \cdot 2 \cdot 2 \cdot 3 \cdot 5 \cdot 7$
    - c. $2 \cdot 2 \cdot 3 \cdot 3 \cdot 5 \cdot 5 \cdot 11$
    - d. $2 \cdot 3 \cdot 5 \cdot 10 \cdot 11$

2. Determine the prime factorization of 990.
    - a. $2 \cdot 3 \cdot 3 \cdot 3 \cdot 5 \cdot 7$
    - b. $3 \cdot 3 \cdot 5 \cdot 5 \cdot 7 \cdot 10$
    - c. $2 \cdot 2 \cdot 3 \cdot 3 \cdot 5 \cdot 5 \cdot 11$
    - d. $2 \cdot 3 \cdot 3 \cdot 5 \cdot 11$

3. Write 700 as a product of prime numbers.

4. Write 525 as a product of prime numbers.

5. Write 1260 as a product of prime numbers.

6. Determine the prime factorization of 2200.
    a. $2 \cdot 2 \cdot 2 \cdot 3 \cdot 5 \cdot 5 \cdot 7$ b. $2 \cdot 2 \cdot 2 \cdot 2 \cdot 5 \cdot 11$
    c. $2 \cdot 2 \cdot 2 \cdot 5 \cdot 5 \cdot 11$ d. $2 \cdot 2 \cdot 5 \cdot 7 \cdot 10$

7. Determine the prime factorization of 630.
    a. $2 \cdot 3 \cdot 3 \cdot 3 \cdot 5 \cdot 11$ b. $2 \cdot 3 \cdot 3 \cdot 5 \cdot 7$
    c. $2 \cdot 2 \cdot 2 \cdot 5 \cdot 5 \cdot 11$ d. $2 \cdot 2 \cdot 5 \cdot 7 \cdot 10$

8. Determine the prime factorization of 440.
    a. $2 \cdot 2 \cdot 2 \cdot 5 \cdot 11$ b. $2 \cdot 2 \cdot 2 \cdot 2 \cdot 5 \cdot 5 \cdot 11$
    c. $2 \cdot 2 \cdot 5 \cdot 5 \cdot 7 \cdot 10$ d. $2 \cdot 2 \cdot 2 \cdot 3 \cdot 5 \cdot 7$

9. Write 4200 as a product of prime numbers.

10. Write 9900 as a product of prime numbers.

11. Write 420 as a product of prime numbers.

12. Determine the prime factorization of 3150.
    a. $2 \cdot 3 \cdot 3 \cdot 3 \cdot 5 \cdot 5 \cdot 11$ b. $3 \cdot 3 \cdot 5 \cdot 10 \cdot 11$
    c. $2 \cdot 2 \cdot 3 \cdot 3 \cdot 5 \cdot 7$ d. $2 \cdot 3 \cdot 3 \cdot 5 \cdot 5 \cdot 7$

13. Determine the prime factorization of 1320.
    a. $2 \cdot 2 \cdot 2 \cdot 3 \cdot 3 \cdot 5 \cdot 7$ b. $2 \cdot 2 \cdot 2 \cdot 3 \cdot 5 \cdot 11$
    c. $2 \cdot 2 \cdot 3 \cdot 5 \cdot 5 \cdot 7 \cdot 10$ d. $2 \cdot 2 \cdot 2 \cdot 2 \cdot 3 \cdot 5 \cdot 5 \cdot 11$

14. Determine the prime factorization of 1100.
    a. $2 \cdot 5 \cdot 7 \cdot 10$ b. $2 \cdot 2 \cdot 5 \cdot 5 \cdot 11$
    c. $2 \cdot 2 \cdot 3 \cdot 5 \cdot 5 \cdot 7$ d. $2 \cdot 2 \cdot 2 \cdot 5 \cdot 11$

15. Write 440 as a product of prime numbers.

16. Write 19,800 as a product of prime numbers.

17. Write 6300 as a product of prime numbers.

18. Determine the prime factorization of 6600.
    a. $2 \cdot 2 \cdot 2 \cdot 2 \cdot 3 \cdot 5 \cdot 11$ b. $2 \cdot 2 \cdot 2 \cdot 3 \cdot 5 \cdot 5 \cdot 11$
    c. $2 \cdot 2 \cdot 2 \cdot 3 \cdot 3 \cdot 5 \cdot 5 \cdot 7$ d. $2 \cdot 2 \cdot 3 \cdot 5 \cdot 7 \cdot 10$

19. Determine the prime factorization of 1400.
    a. $2 \cdot 2 \cdot 2 \cdot 3 \cdot 5 \cdot 5 \cdot 11$ b. $2 \cdot 2 \cdot 5 \cdot 10 \cdot 11$
    c. $2 \cdot 2 \cdot 2 \cdot 2 \cdot 5 \cdot 7$ d. $2 \cdot 2 \cdot 2 \cdot 5 \cdot 5 \cdot 7$

20. Determine the prime factorization of 330.
        a. $2 \cdot 2 \cdot 3 \cdot 5 \cdot 5 \cdot 11$ b. $2 \cdot 3 \cdot 5 \cdot 11$
        c. $2 \cdot 3 \cdot 3 \cdot 5 \cdot 7$ d. $3 \cdot 5 \cdot 5 \cdot 7 \cdot 10$

21. Determine the prime factorization of 416.
        a. $2 \cdot 2 \cdot 2 \cdot 2 \cdot 13$ b. $2 \cdot 2 \cdot 2 \cdot 2 \cdot 2 \cdot 23$
        c. $2 \cdot 2 \cdot 2 \cdot 2 \cdot 2 \cdot 13$ d. $2 \cdot 2 \cdot 2 \cdot 2 \cdot 23$

22. Determine the prime factorization of 882.
        a. $2 \cdot 2 \cdot 3 \cdot 7 \cdot 7$ b. $2 \cdot 3 \cdot 3 \cdot 7 \cdot 7$
        c. $2 \cdot 2 \cdot 3 \cdot 7$ d. $2 \cdot 3 \cdot 3 \cdot 3 \cdot 7$

23. Determine the prime factorization of 210.
        a. $2 \cdot 2 \cdot 3 \cdot 7$ b. $3 \cdot 5 \cdot 7 \cdot 4$
        c. $2 \cdot 3 \cdot 5 \cdot 7 \cdot 9$ d. $2 \cdot 3 \cdot 5 \cdot 7$

24. Determine the prime factorization of 594.
        a. $2 \cdot 3 \cdot 9 \cdot 11$ b. $3 \cdot 3 \cdot 3 \cdot 22$
        c. $3 \cdot 11 \cdot 13$ d. $2 \cdot 3 \cdot 3 \cdot 3 \cdot 11$

25. Write 1716 as a product of prime numbers.

26. Write 380 as a product of prime numbers.

27. Write 1460 as a product of prime numbers.

28. Write 1000 as a product of prime numbers.

29. Write 777 as a product of prime numbers.

30. Write 5280 as a product of prime numbers.

LESSON 12 – Multiples

Example: List the first ten multiples of 2. It is the 2's times table.

$$2, 4, 6, 8, 10, 12, 14, 16, 18, 20$$

Practice problems on multiples.

1. List the first five multiples of 6.
   - a. 6, 12, 18, 24, 30
   - b. 30, 35, 40, 45, 50
   - c. 6, 7, 8, 9, 10
   - d. 0, 1, 6, 12, 18

2. List the first five multiples of 5.
   - a. 0, 1, 5, 10, 15
   - b. 5, 6, 7, 8, 9
   - c. 5, 10, 15, 20, 25
   - d. 25, 30, 35, 40, 45

3. List the first five multiples of 4.
   - a. 4, 8, 12, 16, 20
   - b. 20, 25, 30, 35, 40
   - c. 0, 1, 4, 8, 12
   - d. 4, 5, 6, 7, 8

4. List the first five multiples of 3.
   - a. 3, 6, 9, 12, 15
   - b. 0, 1, 3, 6, 9
   - c. 15, 20, 25, 30, 35
   - d. 3, 4, 5, 6, 7

5. List the first five multiples of 7.
   - a. 7, 14, 21, 28, 35
   - b. 0, 1, 7, 14, 21
   - c. 7, 8, 9, 10, 11
   - d. 35, 40, 45, 50, 55

6. List the first five multiples of 9.
   - a. 45, 50, 55, 60, 65
   - b. 9, 18, 27, 36, 45
   - c. 0, 1, 9, 18, 27
   - d. 9, 10, 11, 12, 13

7. List the first five multiples of 8.
   - a. 8, 9, 10, 11, 12
   - b. 40, 45, 50, 55, 60
   - c. 8, 16, 24, 32, 40
   - d. 0, 1, 8, 16, 24

8. List the first six multiples of 10.

9. List the first five multiples of 12.

10. List the first nine multiples of 11.

LESSON 13 – Least Common Multiple

There are two distinct ways to find the least common multiple (LCM). The LCM means the smallest number that is divisible by all the numbers.

The first way to find the LCM is to list the sets of multiples of all numbers. Use 12 and 16 as an example.

Multiples of 12: {12, 24, 36, 48, 60, 72, 84, 96, ...}
Multiples of 16: {16, 32, 48, 64, 80, 96, ...}

Upon inspection, the first two common multiples of 12 and 16 are 48 and 96. The smallest of these is 48.

The other way to find the LCM is (1) prime factor each number, (2) select every prime factor raised to the highest power and (3) find their product. Again, using 12 and 16.

Step 1: 12 prime factors into 2^2 x 3
16 prime factors into 24

Step 2: 2^4 and 3

Step 3: 2^4 x 3 = 48, the least common multiple

Now try these for good luck.

1. Find the least common multiple of 8 and 4.
    a. 32   b. 8   c. 12   d. 4

2. Find the least common multiple of 6 and 2.
    a. 6   b. 12   c. 8   d. 2

3. What is the least common multiple of 2 and 8?

4. Find the least common multiple of 6, 2, and 12.

5. Find the least common multiple of 2, 10, and 3.

6. Find the least common multiple of 3, 6, and 8.

7. Find the least common multiple of 8, 2, and 9.

8. What is the least common multiple of 2 and 10?

9. What is the least common multiple of 9 and 6?

10. Find the least common multiple of 10 and 8.
    a. 80   b. 2   c. 40   d. 18

11. What is the least common multiple of 6 and 8?

12. Find the least common multiple of 2, 6, and 12.

13. What is the least common multiple of 6 and 9?

14. Find the least common multiple of 6 and 8.
    a. 48   b. 2   c. 24   d. 14

15. Find the least common multiple of 2, 4, and 3.

16. What is the least common multiple of 10 and 5?

17. Find the least common multiple of 8 and 12.
    a. 4   b. 96   c. 24   d. 20

18. Find the least common multiple of 10, 6, and 4.

19. Find the least common multiple of 2, 3, and 5.

20. Find the least common multiple of 3, 5, and 9.

21. What is the least common multiple of 4, 6, and 8?

22. Find the least common multiple of 8, 10, and 12.

23. Find the least common multiple of 24, 12, and 6.

24. What is the least common multiple of 5, 10, and 25?

25. Find the least common multiple of 6, 8, and 9.

26. Find the least common multiple of 2, 5, and 10.

27. Find the least common multiple of 2, 7, and 8.

28. Find the least common multiple of 5, 10, and 20.

29. Find the least common multiple of 6, 9, and 15.

30. Find the least common multiple of 12, 20, and 30.

LESSON 14 – Exponents

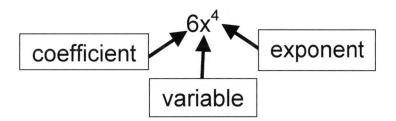

Example 1: $x^3 = x \cdot x \cdot x$

Example 2: $4^3 = 4 \cdot 4 \cdot 4 = 64$

Example 3: $a^2 = a \cdot a$

Example 4: $3b^3 = 3 \cdot b \cdot b \cdot b$

Example 5: $2^5 = 2 \cdot 2 \cdot 2 \cdot 2 \cdot 2 = 32$

Practice on exponents. Evaluate the following.

1. 2^4 a. 17   b. 16   c. 15   d. 18

2. 3^4 a. 12   b. 81   c. 243   d. 64

3. 2^3

4. 5^2

5. 4^2

6. 4^3

7. 5^3

8. 7^2

9. 3^3 a. 26   b. 29   c. 28   d. 27

10. 3^2 a. 9   b. 6   c. 27   d. 8

11. 2^3 a. 16   b. 9   c. 8   d. 6

12. 6^2

13. 8^2

14. 7^3

15. 10^2

16. 9^2

17. 4^4

18. 2^6

19. 8^3

20. 9^3

21. 10^3

22. 5^4

23. 11^2

24. 15^2

25. 12^2

26. 13^2

Expand the following.

27. $4x^2$

28. $16a^3$

29. $(4x)^2$

30. $20b^3$

31. $9c^2$

32. $10a^2b^2$

33. $4x^2yz$

34. $16x^4$

LESSON 15 – Greatest Common Factor

The Greatest Common Factor (GCF) of two numbers is the greatest whole number that is a factor of each number. To find the GCF, find the prime factorization of each number and then find the greatest power of each factor common to BOTH numbers. Their product is the GCF.

Example 1:   Find the GCF of $\frac{16}{72}$.

$$16 = 2 \cdot 2 \cdot 2 \cdot 2 = \boxed{2^3} \cdot 2$$
$$72 = 2 \cdot 2 \cdot 2 \cdot 3 \cdot 3 = \boxed{2^3} \cdot 3^2$$

2^3 is common to both numbers. Therefore the GCF is $2^3 = 8$.

Example 2:   Find the GCF of $\frac{49}{63}$.

$$49 = \boxed{7} \cdot 7$$
$$63 = 7 \cdot 3 \cdot 3 = \boxed{7} \cdot 3^2$$

7 is common to both numbers. Therefore the GCF is 7.

Example 3:   Find the GCF of $\frac{144}{256}$.

$$144 = 2 \cdot 2 \cdot 2 \cdot 2 \cdot 3 \cdot 3 = \boxed{2^4} \cdot 3^2$$
$$256 = 2 \cdot 2 \cdot 2 \cdot 2 \cdot 2 \cdot 2 \cdot 2 \cdot 2 = \boxed{2^4} \cdot 2^4$$

2^4 is common to both numbers. Therefore the GCF is $2^4 = 16$.

Example 4:   Find the GCF of 36 and 78.

$$36 = 2 \cdot 2 \cdot 3 \cdot 3 = \boxed{2 \cdot 3} \cdot 2 \cdot 3$$
$$78 = \boxed{2 \cdot 3} \cdot 13$$

$2 \cdot 3$ is common to both numbers. Therefore the GCF is $2 \cdot 3 = 6$.

Example 5:   Find the GCF of 48 and 120.

$$48 = 2 \cdot 2 \cdot 2 \cdot 2 \cdot 3 = \boxed{2^3 \cdot 3} \cdot 2$$
$$120 = 2 \cdot 2 \cdot 2 \cdot 3 \cdot 5 = \boxed{2^3 \cdot 3} \cdot 5$$

$2^3 \cdot 3$ is common to both numbers.
Therefore the GCF is $2^3 \cdot 3 = 24$.

Practice problems on finding the GCF. Find the GCF for the following.

1. 24 and 32

2. 10 and 15

3. 18 and 45

4. 16 and 64

5. 24 and 42

6. 32 and 80

7. 60 and 108

8. 112 and 128

9. 342 and 380

10. 210 and 252

11. 308 and 418

12. 144 and 300

13. 30 and 48

14. 36 and 150

15. 216 and 254

16. 48 and 72

17. 72 and 120

18. 60 and 108

19. 66 and 90

20. 150 and 480

LESSON 16 – Reducing Fractions

Now we can use the LCM and GCF in working with fractions. To reduce a fraction, find the GCF for both numerator and denominator simultaneously. Divide the GCF into both top and bottom of the fraction. The result is a fraction in lowest (reduced) form. Some examples are:

Example 1: $\dfrac{36}{48} = \dfrac{2 \cdot 2 \cdot 3 \cdot 3}{2 \cdot 2 \cdot 2 \cdot 2 \cdot 3 \cdot 3} = \dfrac{2^2 \cdot 3}{2^2 \cdot 3} \cdot \dfrac{3}{2^2} = \dfrac{\mathbf{12} \cdot 3}{\mathbf{12} \cdot 4} = \dfrac{3}{4}$

Example 2: $\dfrac{40}{50} = \dfrac{2 \cdot 2 \cdot 2 \cdot 5}{2 \cdot 5 \cdot 5} = \dfrac{2 \cdot 5}{2 \cdot 5} \cdot \dfrac{2^2}{5} = \dfrac{\mathbf{10} \cdot 4}{\mathbf{10} \cdot 5} = \dfrac{4}{5}$

Example 3: $\dfrac{2 \cdot 3 \cdot 3}{2 \cdot 2 \cdot 2 \cdot 2 \cdot 2 \cdot 2} = \dfrac{2}{2} \cdot \dfrac{3^2}{2^5} = \dfrac{\mathbf{2} \cdot 9}{\mathbf{2} \cdot 32} = \dfrac{9}{32}$

Example 4: $\dfrac{27}{42} = \dfrac{3 \cdot 3 \cdot 3}{2 \cdot 3 \cdot 7} = \dfrac{3}{3} \cdot \dfrac{3^2}{2 \cdot 7} = \dfrac{\mathbf{3} \cdot 9}{\mathbf{3} \cdot 14} = \dfrac{9}{14}$

Example 5: $\dfrac{12}{60} = \dfrac{2 \cdot 2 \cdot 3}{2 \cdot 2 \cdot 3 \cdot 5} = \dfrac{2^2 \cdot 3}{2^2 \cdot 3} \cdot \dfrac{1}{5} = \dfrac{\mathbf{12} \cdot 1}{\mathbf{12} \cdot 5} = \dfrac{1}{5}$

Example 6: $\dfrac{15}{36} = \dfrac{3 \cdot 5}{2 \cdot 2 \cdot 3 \cdot 3} = \dfrac{3}{3} \cdot \dfrac{5}{2^2 \cdot 3} = \dfrac{\mathbf{3} \cdot 5}{\mathbf{3} \cdot 12} = \dfrac{5}{12}$

Example 7: $\dfrac{56}{91} = \dfrac{2 \cdot 2 \cdot 2 \cdot 7}{7 \cdot 13} = \dfrac{7}{7} \cdot \dfrac{2^3}{13} = \dfrac{\mathbf{7} \cdot 8}{\mathbf{7} \cdot 13} = \dfrac{8}{13}$

Practice on reducing fractions. Reduce to lowest terms.

1. $\dfrac{27}{45}$ a. $\frac{3}{5}$ b. $\frac{2}{9}$ c. $\frac{3}{4}$ d. $\frac{9}{15}$

2. $\dfrac{60}{80}$ a. $\frac{15}{20}$ b. $\frac{4}{5}$ c. $\frac{3}{4}$ d. $\frac{6}{8}$

3. $\dfrac{12}{30}$ a. $\frac{1}{2}$ b. $\frac{4}{10}$ c. $\frac{2}{5}$ d. $\frac{4}{7}$

4. $\dfrac{36}{48}$ a. $\frac{5}{36}$ b. $\frac{3}{4}$ c. $\frac{6}{7}$ d. $\frac{9}{12}$

5. $\frac{48}{60}$ a. $\frac{8}{9}$ b. $\frac{1}{8}$ c. $\frac{4}{5}$ d. $\frac{16}{20}$

6. $\frac{16}{24}$ a. $\frac{4}{5}$ b. $\frac{2}{3}$ c. $\frac{4}{6}$ d. $\frac{1}{4}$

7. $\frac{18}{36}$ a. $\frac{6}{12}$ b. $\frac{1}{2}$ c. $\frac{5}{7}$ d. $\frac{5}{18}$

8. $\frac{16}{20}$ a. $\frac{4}{5}$ b. $\frac{16}{20}$ c. $\frac{3}{32}$ d. $\frac{8}{9}$

9. $\frac{18}{72}$ a. $\frac{1}{5}$ b. $\frac{1}{6}$ c. $\frac{1}{4}$ d. $\frac{5}{18}$

10. $\frac{108}{144}$ a. $\frac{5}{36}$ b. $\frac{3}{4}$ c. $\frac{9}{12}$ d. $\frac{6}{7}$

11. $\frac{48}{54}$ a. $\frac{4}{5}$ b. $\frac{1}{8}$ c. $\frac{8}{9}$ d. $\frac{16}{20}$

12. $\frac{24}{40}$

13. $\frac{60}{72}$

14. $\frac{28}{70}$

15. $\frac{24}{32}$

16. $\frac{18}{27}$ a. $\frac{2}{9}$ b. $\frac{2}{3}$ c. $\frac{5}{6}$ d. $\frac{6}{9}$

17. $\frac{24}{40}$ a. $\frac{5}{7}$ b. $\frac{6}{10}$ c. $\frac{3}{5}$ d. $\frac{1}{4}$

18. $\frac{32}{80}$ a. $\frac{3}{16}$ b. $\frac{8}{20}$ c. $\frac{2}{3}$ d. $\frac{2}{5}$

19. $\frac{18}{30}$ a. $\frac{5}{7}$ b. $\frac{3}{5}$ c. $\frac{1}{3}$ d. $\frac{6}{10}$

20. $\frac{65}{100}$ 21. $\frac{25}{75}$

22. $\frac{16}{40}$ 23. $\frac{66}{99}$

24. $\frac{28}{32}$ 25. $\frac{15}{50}$

26. $\frac{36}{54}$ 27. $\frac{15}{40}$

28. $\frac{20}{48}$

LESSON 17 - Reciprocals

Two numbers, $\frac{a}{b}$ and $\frac{b}{a}$, whose product is 1 are reciprocals.

$$\frac{6}{11} \times \boxed{n} = 1 \quad \rightarrow \quad n = \frac{11}{6}$$

$$7 \times \boxed{n} = 1 \quad \rightarrow \quad n = \frac{1}{7}$$

Another name for the reciprocal is **multiplicative inverse**.

The multiplicative inverse of 5 is $\frac{1}{5}$.

The multiplicative inverse of $-\frac{1}{6}$ is -6.

Practice on reciprocals.

1. a. What is the reciprocal of –9?
    b. What is the reciprocal of $\frac{1}{9}$?

2. Which real number does not have a reciprocal and why?

3. What is the multiplicative inverse of $-\frac{1}{7}$?

         a. 7   b. –7   c. $-\frac{1}{7}$ d. $\frac{1}{7}$

4. What is the reciprocal of $\frac{1}{8}$?

         a. $-\frac{1}{8}$ b. $\frac{1}{8}$ c. 8   d. –8

5. What is the multiplicative inverse of –3?

         a. –3   b. $-\frac{1}{3}$ c. $\frac{1}{3}$ d. 3

6. a. What is the multiplicative inverse of $\frac{1}{8}$?
    b. What is the multiplicative inverse of –8?

7. a. What is the reciprocal of $-\frac{1}{2}$?
    b. What is the reciprocal of 2?

8. What is the reciprocal of 8?

         a. 8   b. – 8   c. $-\frac{1}{8}$ d. $\frac{1}{8}$

9. What is the multiplicative inverse of –7?

    a. $\frac{1}{7}$ b. -7 c. $-\frac{1}{7}$ d. 7

10. What is the reciprocal of $-\frac{1}{5}$?

11. What is the multiplicative inverse of 4?

    a. -4 b. $\frac{1}{4}$ c. $-\frac{1}{4}$ d. 4

12. What is the multiplicative inverse of –2?

    a. -2 b. $-\frac{1}{2}$ c. 2   d. $\frac{1}{2}$

13. What is the multiplicative inverse of 10?

    a. 10   b. -10 c. $-\frac{1}{10}$ d. $\frac{1}{10}$

14. What is the reciprocal of $-\frac{3}{7}$?

    a. $\frac{7}{3}$ b. $-\frac{7}{3}$ c. $\frac{3}{7}$ d. $-\frac{3}{7}$

15. What is the reciprocal of $\frac{23}{9}$?

    a. $-\frac{23}{9}$ b. $\frac{23}{9}$ c. $\frac{9}{23}$ d. $-\frac{9}{23}$

16. What is the multiplicative inverse of $\frac{17}{35}$?

    a. $\frac{17}{35}$ b. $-\frac{17}{35}$ c. $\frac{35}{17}$ d. $-\frac{35}{17}$

17. What is the reciprocal of $-\frac{2}{3}$?

    a. $\frac{2}{3}$ b. $-\frac{2}{3}$ c. $\frac{3}{2}$ d. $-\frac{3}{2}$

18. What is the multiplicative inverse of $\frac{4}{9}$?

    a. $\frac{9}{4}$ b. $-\frac{9}{4}$ c. $\frac{4}{9}$ d. $-\frac{4}{9}$

19. What is the reciprocal of $-\frac{6}{11}$?

    a. $-\frac{11}{6}$ b. $\frac{11}{6}$ c. $-\frac{6}{11}$ d. $\frac{6}{11}$

20. What is the reciprocal of 13?

    a. 13   b. $-\frac{1}{13}$ c. -13 d. $\frac{1}{13}$

LESSON 18 – Multiplying and Dividing Fractions

The rule for multiplication of fractions is $\dfrac{a}{b} \cdot \dfrac{c}{d} = \dfrac{ac}{bd}$, where "b" and "d" are not equal to zero.

Example 1: $\quad \dfrac{2}{3} \times \dfrac{4}{5} = \dfrac{2 \times 4}{3 \times 5} = \dfrac{8}{15}$

Example 2: $\quad \dfrac{7}{8} \cdot \dfrac{4}{7} = \dfrac{{}^{1}\cancel{7} \cdot 4}{8 \cdot {}_{1}\cancel{7}} = \dfrac{1 \cdot 4}{8 \cdot 1} = \dfrac{4}{8} = \dfrac{1}{2}$

$\left.\begin{array}{l} \\ \\ \\ \\ \end{array}\right\}$
1. Remember to cancel when the same number is in the top and bottom.
2. Also, reduce fraction to lowest terms.

Example 3: $\quad \dfrac{5}{12} \cdot \dfrac{4}{10} = \dfrac{{}^{1}\cancel{5} \cdot {}^{1}\cancel{4}}{{}_{3}\cancel{12} \cdot {}_{2}\cancel{10}} = \dfrac{1}{3 \cdot 2} = \dfrac{1}{6}$

Example 4: $\quad \dfrac{4}{9} \times \dfrac{10}{15} = \dfrac{4 \times {}^{2}\cancel{10}}{9 \times {}_{3}\cancel{15}} = \dfrac{4 \times 2}{9 \times 3} = \dfrac{8}{27}$

The rule for division of fractions is to multiply by the reciprocal of the **divisor**.

$$\dfrac{a}{b} \div \dfrac{c}{d} = \dfrac{a}{b} \cdot \dfrac{d}{c} = \dfrac{ad}{bc} \qquad \text{where "b", "c", and "d"} \neq 0$$

divisor   reciprocal

Example 5: $\quad \dfrac{4}{7} \div \dfrac{9}{20} = \dfrac{4}{7} \cdot \dfrac{20}{9} = \dfrac{4 \cdot 20}{7 \cdot 9} = \dfrac{80}{63} = 1\dfrac{17}{63}$

Example 6: $\quad \dfrac{3}{2} \div \dfrac{3}{8} = \dfrac{3}{2} \cdot \dfrac{8}{3} = \dfrac{{}^{1}\cancel{3} \cdot {}^{4}\cancel{8}}{{}_{1}\cancel{2} \cdot {}_{1}\cancel{3}} = \dfrac{1 \cdot 4}{1 \cdot 1} = \dfrac{4}{1} = 4$

Example 7: $\quad \dfrac{3}{4} \div \dfrac{7}{8} = \dfrac{3}{4} \cdot \dfrac{8}{7} = \dfrac{3 \cdot {}^{2}\cancel{8}}{{}_{1}\cancel{4} \cdot 7} = \dfrac{3 \cdot 2}{1 \cdot 7} = \dfrac{6}{7}$

Example 8: $\quad \dfrac{2}{3} \div \dfrac{1}{6} = \dfrac{2}{3} \cdot \dfrac{6}{1} = \dfrac{2 \cdot {}^{2}\cancel{6}}{{}_{1}\cancel{3} \cdot 1} = \dfrac{2 \cdot 2}{1 \cdot 1} = \dfrac{4}{1} = 4$

Practice problems. Multiply and reduce your answer where necessary.

1. $\frac{2}{5} \times \frac{4}{7}$

2. $\frac{2}{7} \times \frac{2}{9}$

3. $4 \times \frac{3}{8}$ a. 4   b. $4\frac{3}{8}$ c. 2   d. $1\frac{1}{2}$

4. $6 \times \frac{5}{9}$ a. 3   b. $5\frac{5}{9}$ c. $3\frac{4}{9}$ d. $3\frac{1}{3}$

5. $\frac{19}{20} \times \frac{4}{19}$

6. $\frac{8}{9} \times \frac{3}{8}$

7. What is the product of $\frac{2}{5}$ and $\frac{2}{11}$?

8. What is the product of $\frac{2}{3}$ and $\frac{2}{5}$?

9. $\frac{5}{6} \cdot \frac{4}{5}$

10. $\frac{6}{7} \cdot \frac{4}{11}$

11. $14 \times \frac{2}{21}$

12. $6 \times \frac{5}{8}$ a. $3\frac{3}{4}$ b. $6\frac{1}{2}$ c. 4   d. $6\frac{5}{8}$

13. $\frac{9}{10} \cdot \frac{2}{9}$

14. What is the product of $\frac{2}{7}$ and $\frac{2}{3}$?

15. What is the product of $\frac{2}{5}$ and $\frac{4}{11}$?

16. How many $\frac{1}{5}$'s are in $\frac{8}{10}$? $\left(\frac{8}{10} \div \frac{1}{5} \right)$

17. How many $\frac{3}{4}$'s are in $\frac{1}{12}$? $\left(\frac{1}{12} \div \frac{3}{4} \right)$

        a. $\frac{1}{9}$ b. 16   c. $\frac{1}{16}$ d. 9

18. How many $\frac{4}{5}$'s are in $\frac{1}{20}$? $\left(\frac{1}{20} \div \frac{4}{5} \right)$

        a. $\frac{1}{16}$ b. 25   c. 16   d. $\frac{1}{25}$

19. How many $\frac{1}{5}$'s are in $\frac{9}{20}$? $\left(\frac{9}{20} \div \frac{1}{5} \right)$

20. How many $\frac{2}{3}$'s are in $\frac{1}{6}$? $\left(\frac{1}{6} \div \frac{2}{3} \right)$

        a. 4   b. $\frac{1}{4}$ c. 9   d. $\frac{1}{9}$

21. How many $\frac{1}{3}$'s are in $\frac{5}{6}$? $\left(\frac{5}{6} \div \frac{1}{3} \right)$

22. How many $\frac{1}{5}$'s are in $\frac{11}{15}$? $\left(\frac{11}{15} \div \frac{1}{5} \right)$

23. How many $\frac{1}{7}$'s are in $\frac{11}{14}$? $\left(\frac{11}{14} \div \frac{1}{7} \right)$

24. $\dfrac{3}{8} \div \dfrac{1}{4}$

25. $\dfrac{5}{12} \div \dfrac{5}{4}$

26. $\dfrac{7}{8} \times \dfrac{4}{3}$

27. $\dfrac{24}{29} \div \dfrac{6}{29}$

28. $\dfrac{14}{15} \cdot \dfrac{30}{28}$

29. $\dfrac{9}{16} \div \dfrac{3}{4}$

30. $\dfrac{27}{33} \cdot \dfrac{22}{3}$

LESSON 19 – Multiple Fractional Factors

Here we have more than two fractions to multiply.

Example 1: $\dfrac{2}{3} \times \dfrac{3}{4} \times \dfrac{15}{60} = \dfrac{{}^{1}2 \times {}^{1}3 \times {}^{1}\cancel{15}}{{}_{1}3 \times {}_{2}4 \times {}_{4}\cancel{60}} = \dfrac{1 \times 1 \times 1}{1 \times 2 \times 4} = \dfrac{1}{8}$

Cancel the 3's, reduce 2 and 4 by 2, finally reduce 15 and 60 by 15. If necessary put 15 and 60 in prime factor form. 15 = 3 × 5 and 60 = 2 × 2 × 3 × 5.

Prime factor form:

$$\frac{2}{3} \times \frac{3}{4} \times \frac{15}{60} = \frac{2}{3} \cdot \frac{3}{2 \times 2} \cdot \frac{3 \times 5}{2 \times 2 \times 3 \times 5} =$$

$$\frac{{}^{1}2}{{}_{1}3} \cdot \frac{{}^{1}3}{{}_{1}2 \times 2} \cdot \frac{{}^{1}(\cancel{3 \times 5})}{2 \times 2 \times {}_{1}(\cancel{3 \times 5})} = \frac{1 \times 1 \times 1}{1 \times 2 \times 4} = \frac{1}{8}$$

Example 2: $\dfrac{5}{8} \cdot \dfrac{3}{6} \cdot \dfrac{25}{70} = \dfrac{5}{8} \cdot \dfrac{{}^{1}3}{{}_{2}6} \cdot \dfrac{{}^{5}\cancel{25}}{{}_{14}\cancel{70}} = \dfrac{5 \cdot 1 \cdot 5}{8 \cdot 2 \cdot 14} = \dfrac{25}{224}$

Example 3: $\dfrac{1}{4} \times \dfrac{1}{3} \times \dfrac{1}{2} = \dfrac{1 \times 1 \times 1}{4 \times 3 \times 2} = \dfrac{1}{24}$

Example 4: $\dfrac{5}{16} \cdot \dfrac{1}{2} \cdot \dfrac{8}{10} = \dfrac{5}{2 \cdot 2 \cdot 2 \cdot 2} \cdot \dfrac{1}{2} \cdot \dfrac{2 \cdot 2 \cdot 2}{2 \cdot 5} =$

$$\frac{{}^{1}\cancel{5}}{{}_{1}(2 \cdot 2 \cdot 2) \cdot 2} \cdot \frac{1}{2} \cdot \frac{{}^{1}(\cancel{2 \cdot 2 \cdot 2})}{2 \cdot {}_{1}\cancel{5}} = \frac{1 \cdot 1 \cdot 1}{2 \cdot 2 \cdot 2} = \frac{1}{8}$$

Example 5: $\dfrac{4}{5} \times \dfrac{3}{8} \times \dfrac{20}{55} = \dfrac{{}^{1}\cancel{4}}{{}_{1}\cancel{5}} \times \dfrac{3}{{}_{12}\cancel{8}} \times \dfrac{{}^{2}{}^{4}\cancel{20}}{55} = \dfrac{1 \times 3 \times 2}{1 \times 1 \times 55} = \dfrac{6}{55}$

Example 6: $\dfrac{2}{3} \cdot \dfrac{4}{7} \cdot \dfrac{28}{36} = \dfrac{2}{3} \cdot \dfrac{{}^{1}\cancel{4}}{{}_{1}\cancel{7}} \cdot \dfrac{{}^{4}\cancel{28}}{{}_{9}\cancel{36}} = \dfrac{2 \cdot 1 \cdot 4}{3 \cdot 1 \cdot 9} = \dfrac{8}{27}$

Practice on Multiple Fractional Factors.

1. $\frac{2}{5} \times \frac{1}{7} \times \frac{14}{105}$ a. $\frac{20}{525}$ b. $\frac{4}{525}$ c. $\frac{4}{15}$ d. $\frac{4}{105}$

2. $\frac{3}{5} \cdot \frac{3}{4} \cdot \frac{15}{80}$ a. $\frac{135}{40}$ b. $\frac{27}{16}$ c. $\frac{27}{64}$ d. $\frac{27}{320}$

3. $\frac{1}{7} \times \frac{3}{5} \times \frac{20}{175}$ a. $\frac{84}{1225}$ b. $\frac{12}{1225}$ c. $\frac{12}{35}$ d. $\frac{12}{175}$

4. $\frac{2}{7} \times \frac{1}{7} \times \frac{21}{196}$ a. $\frac{42}{1372}$ b. $\frac{3}{686}$ c. $\frac{3}{98}$ d. $\frac{3}{14}$

5. $\frac{2}{5} \times \frac{3}{4} \times \frac{20}{100}$ a. $\frac{3}{10}$ b. $\frac{3}{50}$ c. $\frac{120}{500}$ d. $\frac{6}{5}$

6. $\frac{1}{5} \cdot \frac{1}{7} \cdot \frac{21}{140}$ a. $\frac{15}{700}$ b. $\frac{3}{700}$ c. $\frac{3}{20}$ d. $\frac{3}{140}$

7. $\frac{3}{7} \times \frac{3}{5} \times \frac{10}{105}$ a. $\frac{126}{735}$ b. $\frac{6}{7}$ c. $\frac{6}{245}$ d. $\frac{6}{35}$

8. $\frac{1}{3} \cdot \frac{1}{4} \cdot \frac{9}{48}$ a. $\frac{1}{64}$ b. $\frac{3}{16}$ c. $\frac{9}{144}$ d. $\frac{3}{64}$

9. $\frac{1}{7} \times \frac{1}{7} \times \frac{14}{147}$ a. $\frac{2}{1029}$ b. $\frac{2}{21}$ c. $\frac{2}{147}$ d. $\frac{14}{1029}$

10. $\frac{2}{7} \times \frac{3}{5} \times \frac{20}{175}$ a. $\frac{24}{1225}$ b. $\frac{168}{1225}$ c. $\frac{24}{175}$ d. $\frac{24}{35}$

11. $\frac{2}{3} \cdot \frac{3}{4} \cdot \frac{6}{36}$ a. $\frac{1}{4}$ b. 1 c. $\frac{36}{108}$ d. $\frac{1}{12}$

12. $\frac{3}{5} \times \frac{1}{7} \times \frac{21}{140}$ a. $\frac{9}{140}$ b. $\frac{9}{700}$ c. $\frac{9}{20}$ d. $\frac{45}{700}$

13. $\frac{5}{8} \times \frac{16}{25} \times \frac{8}{12}$ a. 3 b. $\frac{1}{3}$ c. $\frac{4}{15}$ d. $\frac{14}{3}$

14. $\frac{3}{16} \times \frac{8}{5} \times \frac{12}{3}$ a. $\frac{5}{6}$ b. $\frac{6}{5}$ c. $\frac{7}{8}$ d. $1\frac{2}{5}$

15. $\frac{5}{9} \times \frac{5}{6} \times \frac{2}{5}$ a. $\frac{50}{341}$ b. $\frac{6}{31}$ c. $\frac{4}{25}$ d. $\frac{5}{27}$

LESSON 20 – Changing a Mixed Number to an Improper Fraction

Example 1: Change $4\frac{5}{8}$ to an improper fraction.

Step 1: Multiply the whole number by the denominator of the fraction.

Step 2: Add the numerator of the fraction to the product of step 1.

Step 3: Rewrite the fraction by taking the step 2 answer as the new numerator over the original fraction's denominator.

$$4 \times 8 = 32$$

$$32 + 5 = 37$$

$$4\frac{5}{8} = \frac{37}{8}$$

In short: $\dfrac{(4 \times 8) + 5}{8}$

Example 2: Change $3\frac{2}{5}$ to an improper fraction.

Step 1: Multiply the whole number by the denominator of the fraction.

Step 2: Add the numerator of the fraction to the product of step 1.

Step 3: Rewrite the fraction by taking the step 2 answer as the new numerator over the original fraction's denominator.

$$3 \times 5 = 15$$

$$15 + 2 = 17$$

$$3\frac{2}{5} = \frac{17}{5}$$

In short: $\dfrac{(3 \times 5) + 2}{5}$

Example 3: Change $17\frac{2}{3}$ to an improper fraction.

Step 1: Multiply the whole number by the denominator of the fraction.

Step 2: Add the numerator of the fraction to the product of step 1.

Step 3: Rewrite the fraction by taking the step 2 answer as the new numerator over the original fraction's denominator.

$$17 \cdot 3 = 51$$

$$51 + 2 = 53$$

$$17\frac{2}{5} = \frac{53}{3}$$

In short: $\dfrac{(17 \cdot 3) + 2}{3}$

Now try these for practice. Write as improper fractions.

1. $4\frac{2}{3}$ a. $\frac{12}{3}$ b. $\frac{6}{3}$ c. 14   d. $\frac{14}{3}$

2. $5\frac{1}{2}$ a. $\frac{10}{2}$ b. $\frac{51}{2}$ c. $\frac{11}{2}$ d. $\frac{5}{2}$

3. $2\frac{4}{5}$

4. $1\frac{1}{2}$

5. $4\frac{1}{3}$ a. $\frac{13}{3}$ b. $\frac{41}{3}$ c. $\frac{12}{3}$ d. $\frac{4}{3}$

6. $2\frac{1}{2}$

7. $2\frac{3}{4}$

8. $3\frac{2}{5}$

9. $3\frac{3}{5}$ a. $\frac{18}{5}$ b. $\frac{15}{5}$ c. $\frac{33}{5}$ d. $\frac{9}{5}$

10. $2\frac{1}{3}$ a. $\frac{6}{3}$ b. $\frac{7}{3}$ c. $\frac{2}{3}$ d. 7

11. $5\frac{3}{4}$ a. $\frac{15}{4}$ b. $\frac{53}{4}$ c. $\frac{20}{4}$ d. $\frac{23}{4}$

12. $4\frac{1}{5}$

13. $7\frac{2}{7}$

14. $8\frac{3}{4}$

15. $4\frac{3}{8}$

16. $6\frac{2}{3}$

17. $5\frac{4}{5}$

18. $3\frac{9}{11}$

19. $5\frac{7}{8}$

LESSON 21 – Multiplying Fractions and/or Whole Numbers Together

Remember to cancel when possible (top to bottom or bottom to top).

Example 1: $6 \times \dfrac{1}{2} = \dfrac{6}{1} \times \dfrac{1}{2} = \dfrac{6 \times 1}{1 \times 2} = \dfrac{6}{2} = 3$

Example 2: $5 \cdot \dfrac{2}{3} = \dfrac{5}{1} \cdot \dfrac{2}{3} = \dfrac{5 \cdot 2}{1 \cdot 3} = \dfrac{10}{3} = 3\frac{1}{3}$

Example 3: $\dfrac{2}{3} \cdot \dfrac{5}{7} = \dfrac{2 \cdot 5}{3 \cdot 7} = \dfrac{10}{21}$

Example 4: $\dfrac{4}{5} \times \dfrac{6}{11} = \dfrac{4 \times 6}{5 \times 11} = \dfrac{24}{55}$

Example 5: $\dfrac{2}{3} \times \dfrac{6}{11} = \dfrac{2 \times \overset{2}{\cancel{6}}}{\underset{1}{\cancel{3}} \times 11} = \dfrac{2 \times 2}{1 \times 11} = \dfrac{4}{11}$

Example 6: $\dfrac{5}{8} \cdot \dfrac{4}{5} \cdot \dfrac{7}{12} = \dfrac{\overset{1}{\cancel{5}} \cdot \overset{1}{\cancel{4}} \cdot 7}{\underset{2}{\cancel{8}} \cdot \underset{1}{\cancel{5}} \cdot 12} = \dfrac{1 \cdot 1 \cdot 7}{2 \cdot 1 \cdot 12} = \dfrac{7}{24}$

Example 7: $6 \cdot \dfrac{4}{3} \cdot \dfrac{7}{8} = \dfrac{6}{1} \cdot \dfrac{4}{3} \cdot \dfrac{7}{8} = \dfrac{2 \cdot 3}{1} \cdot \dfrac{(2 \cdot 2)}{3} \cdot \dfrac{7}{2 \cdot (2 \cdot 2)} =$

$\dfrac{\overset{1}{\cancel{2}} \cdot \overset{1}{\cancel{3}}}{1} \cdot \dfrac{\overset{1}{\cancel{(2 \cdot 2)}}}{\underset{1}{\cancel{3}}} \cdot \dfrac{7}{\underset{1}{\cancel{2}} \cdot \underset{1}{\cancel{(2 \cdot 2)}}} = \dfrac{1 \cdot 1 \cdot 7}{1 \cdot 1 \cdot 1} = \dfrac{7}{1} = 7$

Now try these problems. Multiply and reduce your answer where necessary.

1. $6 \times \dfrac{5}{9}$ a. $6\frac{1}{3}$ b. $6\frac{5}{9}$ c. 3   d. $3\frac{1}{3}$

2. $\dfrac{8}{9} \cdot \dfrac{2}{3}$

3. $\dfrac{7}{8} \times \dfrac{4}{7}$

4. What is the product of $\frac{2}{3}$ and $\frac{5}{7}$?

5. What is the product of $\frac{2}{5}$ and $\frac{2}{3}$?

44

6. $\dfrac{11}{12} \times \dfrac{4}{11}$

7. $\dfrac{7}{9} \times \dfrac{4}{5}$

8. $\dfrac{3}{5} \cdot \dfrac{9}{11}$

9. $\dfrac{2}{3} \times \dfrac{5}{11}$

10. $6 \times \dfrac{4}{9}$ a. $6\frac{4}{9}$ b. $4\frac{1}{9}$ c. 3   d. $2\frac{2}{3}$

11. $9 \cdot \dfrac{1}{6}$ a. $2\frac{1}{2}$ b. 2   c. $1\frac{1}{2}$ d. $9\frac{1}{6}$

12. $\dfrac{8}{9} \times \dfrac{3}{8}$

13. What is the product of $\frac{2}{5}$ and $\frac{8}{9}$?

14. $\dfrac{11}{12} \cdot \dfrac{3}{11}$

15. What is the product of $\frac{2}{5}$ and $\frac{8}{9}$?

16. What is the product of $\frac{2}{3}$ and $\frac{10}{11}$?

17. What is the product of $\frac{4}{9}$ and $\frac{2}{3}$?

18. What is the product of $\frac{4}{7}$ and $\frac{2}{5}$?

19. What is the product of $\frac{2}{9}$ and $\frac{2}{3}$?

20. $6 \cdot \dfrac{5}{8}$ a. $3\frac{3}{4}$ b. 4   c. $6\frac{5}{8}$ d. $6\frac{1}{8}$

21. $\dfrac{5}{7} \times \dfrac{8}{11}$

22. What is the product of $\frac{4}{5}$ and $\frac{2}{3}$?

LESSON 22 – Dividing Fractions

There are two ways to write division of fractions:

$$\frac{a}{b} \div \frac{c}{d} \quad \text{or} \quad \frac{\dfrac{a}{b}}{\dfrac{c}{d}}$$

Take the reciprocal of the second (or bottom) fraction and multiply times the first (or top) fraction.

Reciprocal: Two numbers, $\frac{a}{b}$ and $\frac{b}{a}$, whose product is 1.

$$\frac{a}{b} \div \frac{c}{d} = \frac{a}{b} \times \frac{d}{c} = \frac{a \times d}{b \times c} = \frac{ad}{bc}$$

$$\underbrace{\phantom{\frac{a}{b} \div \frac{c}{d}}}_{\text{reciprocal}}$$

$$\frac{\dfrac{a}{b} \times \dfrac{d}{c}}{\dfrac{c}{d} \times \dfrac{d}{c}} = \frac{\dfrac{a}{b} \times \dfrac{d}{c}}{1} = \frac{a}{b} \cdot \frac{d}{c} = \frac{a \cdot d}{b \cdot c} = \frac{ad}{bc}$$

Example 1: $\dfrac{4}{5} \div \dfrac{3}{8} = \dfrac{4}{5} \cdot \dfrac{8}{3} = \dfrac{4 \cdot 8}{5 \cdot 3} = \dfrac{32}{15} = 2\frac{2}{15}$

Example 2: $\dfrac{5}{8} \div \dfrac{3}{4} = \dfrac{5}{8} \times \dfrac{4}{3} = \dfrac{5 \times {}^{1}4}{{}_{2}8 \times 3} = \dfrac{5 \times 1}{2 \times 3} = \dfrac{5}{6}$

Example 3: $\dfrac{3}{4} \div 8 = \dfrac{3}{4} \div \dfrac{8}{1} = \dfrac{3}{4} \times \dfrac{1}{8} = \dfrac{3 \times 1}{4 \times 8} = \dfrac{3}{32}$

Example 4: $\dfrac{\dfrac{5}{7}}{\dfrac{3}{8}} = \dfrac{\dfrac{5}{7} \cdot \dfrac{8}{3}}{\dfrac{3}{8} \cdot \dfrac{8}{3}} = \dfrac{\dfrac{40}{21}}{\dfrac{24}{24}} = \dfrac{\dfrac{40}{21}}{1} = \dfrac{40}{21} = 1\frac{19}{21}$

Example 5: $\dfrac{\dfrac{9}{16}}{\dfrac{1}{4}} = \dfrac{\dfrac{9}{16} \times \dfrac{4}{1}}{\dfrac{1}{4} \times \dfrac{4}{1}} = \dfrac{\dfrac{36}{16}}{\dfrac{4}{4}} = \dfrac{\dfrac{36}{16}}{1} = \dfrac{36}{16} = 2\frac{4}{16} = 2\frac{1}{4}$

Try these for practice. Divide and reduce your answer where necessary.

1. $\frac{2}{3} \div 5$

2. $\frac{2}{3} \div \frac{1}{5}$

3. $\frac{2}{5} \div \frac{3}{5}$

4. $10 \div \frac{1}{5}$

5. $\frac{7}{8} \div \frac{3}{8}$

6. $\frac{5}{8} \div \frac{5}{9}$

7. $\frac{3}{4} \div \frac{7}{8}$

8. $\frac{6}{5} \div 10$

9. $\dfrac{\frac{5}{9}}{\frac{1}{7}}$

10. $\dfrac{\frac{2}{3}}{\frac{3}{8}}$

11. $\dfrac{\frac{5}{16}}{\frac{3}{4}}$

12. $\dfrac{\frac{9}{8}}{\frac{8}{9}}$

13. $\dfrac{\frac{9}{8}}{\frac{9}{8}}$

14. $\dfrac{\frac{4}{7}}{\frac{9}{5}}$

15. $\dfrac{\frac{5}{8}}{\frac{3}{4}}$

16. $\dfrac{\frac{3}{8}}{\frac{5}{16}}$

17. $14 \div \dfrac{2}{3}$

18. $\dfrac{\frac{6}{7}}{\frac{15}{19}}$

19. $\dfrac{\frac{12}{13}}{\frac{2}{5}}$

20. $\dfrac{17}{24} \div \dfrac{5}{8}$

LESSON 23 – Multiplying and Dividing Fractions and Mixed Numbers

Remember: You must change mixed numbers to improper fractions before multiplying and/or dividing.

Example 1: $2\frac{3}{4} \div 1\frac{5}{8} = \frac{11}{4} \div \frac{13}{8} = \frac{11}{4} \cdot \frac{8}{13} = \frac{11}{4} \cdot \frac{\overset{2}{\cancel{8}}}{13} = \frac{11 \cdot 2}{1 \cdot 13} = \frac{22}{13} = 1\frac{9}{13}$

Example 2: $\dfrac{4\frac{7}{12}}{3\frac{2}{3}} = \dfrac{\frac{55}{12}}{\frac{11}{3}} = \frac{55}{12} \div \frac{11}{3} = \frac{55}{12} \times \frac{3}{11} = \frac{\overset{5}{\cancel{55}}}{\underset{4}{\cancel{12}}} \times \frac{\overset{1}{\cancel{3}}}{\underset{1}{\cancel{11}}} = \frac{5 \times 1}{4 \times 1} = \frac{5}{4} = 1\frac{1}{4}$

Example 3: $\frac{5}{8} \times \frac{3}{5} \times \frac{4}{9} = \frac{\overset{1}{\cancel{5}} \times \overset{1}{\cancel{3}} \times \overset{1}{\cancel{4}}}{\underset{2}{\cancel{8}} \times \underset{1}{\cancel{5}} \times \underset{3}{\cancel{9}}} = \frac{1 \times 1 \times 1}{2 \times 1 \times 3} = \frac{1}{6}$

Example 4: $2\frac{1}{4} \times \frac{5}{8} = \frac{9}{4} \times \frac{5}{8} = \frac{9 \times 5}{4 \times 8} = \frac{45}{32} = 1\frac{13}{32}$

Example 5: $4\frac{1}{5} \cdot 3\frac{2}{3} = \frac{21}{5} \cdot \frac{11}{3} = \frac{231}{15} = 15\frac{6}{15} = 15\frac{2}{5}$

or $\frac{21}{5} \cdot \frac{11}{3} = \frac{\overset{7}{\cancel{21}}}{5} \cdot \frac{11}{\underset{1}{\cancel{3}}} = \frac{77}{5} = 15\frac{2}{5}$

Example 6: $4\frac{5}{8} \div 2\frac{1}{3} = \frac{37}{8} \div \frac{7}{3} = \frac{37}{8} \times \frac{3}{7} = \frac{111}{56} = 1\frac{55}{56}$

Example 7: $5\frac{7}{8} \cdot 4\frac{1}{4} = \frac{47}{8} \cdot \frac{17}{4} = \frac{799}{32} = 24\frac{31}{32}$

Now try these. Multiply or divide as indicated. Reduce your answer where necessary.

1. $1\frac{1}{9} \div 2\frac{7}{9}$ a. $\frac{2}{5}$ b. $1\frac{4}{5}$ c. $8\frac{1}{10}$ d. $\frac{11}{18}$

2. $4\frac{1}{2} \div 8\frac{1}{3}$ a. $\frac{27}{50}$ b. $\frac{18}{25}$ c. $2\frac{8}{9}$ d. $\frac{3}{5}$

3. $\frac{6}{7} \times \frac{3}{5} \times \frac{6}{35}$ a. none of these   b. $\frac{108}{35}$ c. $\frac{108}{1225}$ d. $\frac{108}{175}$

4. $\frac{4}{5} \cdot \frac{1}{7} \cdot \frac{1}{8}$ a. $\frac{1}{2}$ b. none of these   c. $\frac{1}{14}$ d. $\frac{1}{70}$

5. $\dfrac{1\frac{2}{3}}{2\frac{1}{5}}$

6. $\dfrac{13\frac{1}{2}}{6\frac{1}{2}}$

7. $2\frac{2}{9} \div 3\frac{4}{7}$

8. $5\frac{1}{6} \cdot 1\frac{1}{5} \cdot 2\frac{1}{2}$

9. $\dfrac{1}{3} \times \dfrac{3}{8} \times \dfrac{1}{6}$

10. $\dfrac{2}{3} \times \dfrac{3}{5} \times \dfrac{3}{4}$

11. $6\frac{6}{7} \cdot 1\frac{4}{7} \cdot 3\frac{1}{2}$

12. $6\frac{1}{2} \times 1\frac{4}{7} \times 4\frac{6}{7}$

13. $1\frac{2}{3} \times 5\frac{2}{3} \times 3\frac{1}{3}$

14. $2\frac{2}{5} \cdot 5\frac{1}{2} \cdot 3\frac{1}{4}$

15. $\dfrac{3\frac{1}{2}}{4\frac{2}{3}}$

16. $\dfrac{1\frac{4}{7}}{4\frac{3}{4}}$

17. $\dfrac{6}{7} \times \dfrac{7}{5} \times \dfrac{6}{35}$ a. $\frac{36}{35}$ b. $\frac{36}{175}$ c. $\frac{252}{1715}$ d. none of these

18. $6\frac{2}{5} \div 1\frac{7}{9}$ a. $2\frac{1}{4}$ b. $1\frac{13}{32}$ c. $3\frac{3}{5}$ d. $2\frac{5}{14}$

19. $3\frac{3}{4} \cdot 4\frac{7}{8}$

20. $10\frac{2}{5} \div 1\frac{11}{15}$

LESSON 24 – Adding Mixed Numbers

Example 1: $6\frac{2}{3} + 4\frac{3}{4}$

Step 1: Convert the fractions $\frac{2}{3}$ and $\frac{3}{4}$ to the same denominator. Get the LCM for 3 and 4, which is 12.

$$6\frac{2}{3} = 6\frac{8}{12}$$
$$+ 4\frac{3}{4} = 4\frac{9}{12}$$

$$\begin{array}{rcl} \frac{2}{3} &=& \frac{x}{12} \\ 3x &=& 24 \\ x &=& 8 \\ \frac{2}{3} &=& \frac{8}{12} \end{array}$$

Step 2: Add the whole numbers and the fractions.

$$6\frac{2}{3} = 6\frac{8}{12}$$
$$+ 4\frac{3}{4} = 4\frac{9}{12}$$
$$\overline{10\frac{17}{12}}$$

OOPS! An improper fraction!

$$\begin{array}{rcl} \frac{3}{4} &=& \frac{x}{12} \\ 4x &=& 36 \\ x &=& 9 \\ \frac{3}{4} &=& \frac{9}{12} \end{array}$$

Step 3: Simplify the improper fraction and add to the whole number.

$$10\frac{17}{12} = 10 + \frac{17}{12} = 10 + 1\frac{5}{12} = 11\frac{5}{12}$$

Example 2: $4\frac{2}{5} + 5\frac{3}{10}$

$$4\frac{2}{5} = 4\frac{4}{10}$$
$$+ 5\frac{3}{10} = 5\frac{3}{10}$$
$$\overline{9\frac{7}{10}}$$

Example 3: $5\frac{7}{8} + 4\frac{2}{3}$

$$5\frac{7}{8} = 5\frac{21}{24}$$
$$+ 4\frac{2}{3} = 4\frac{16}{24}$$
$$\overline{9\frac{37}{24}} = 10\frac{13}{24}$$

Example 4: $2\frac{1}{2} + 3\frac{2}{3}$

$$2\frac{1}{2} = 2\frac{3}{6}$$
$$+ 3\frac{2}{3} = 3\frac{4}{6}$$
$$\overline{5\frac{7}{6}} = 6\frac{1}{6}$$

Example 5: $8\frac{4}{5} + 6\frac{2}{3}$

$$8\frac{4}{5} = 8\frac{12}{15}$$
$$+ 6\frac{2}{3} = 6\frac{10}{15}$$
$$\overline{14\frac{22}{15}} = 15\frac{7}{15}$$

Now try these. Remember to reduce your answer where necessary.

1. $4\frac{2}{3} + 8\frac{3}{5}$

2. $9\frac{1}{3} + 3\frac{3}{10}$

3. $6\frac{2}{3} + 5\frac{3}{4}$

4. $2\frac{1}{3} + 1\frac{3}{8}$

5. $7\frac{1}{3} + 8\frac{3}{5}$

6. $5\frac{1}{5} + 2\frac{1}{2}$ a. $7\frac{2}{7}$ b. 8 c. $7\frac{7}{10}$ d. $5\frac{1}{7}$

7. $7\frac{9}{10} + 1\frac{1}{9}$ a. $8\frac{10}{19}$ b. $8\frac{17}{19}$ c. $9\frac{1}{90}$ d. 9

8. $8\frac{3}{7} + 1\frac{4}{5}$ a. $10\frac{8}{35}$ b. 10 c. $9\frac{7}{12}$ d. $8\frac{2}{3}$

9. $5\frac{2}{3} + 9\frac{3}{4}$

10. $2\frac{2}{3} + 4\frac{3}{5}$

11. $5\frac{1}{5} + 1\frac{1}{6}$ a. $5\frac{7}{8}$ b. 7 c. $6\frac{11}{30}$ d. $6\frac{3}{16}$

12. $2\frac{2}{9} + 1\frac{3}{7}$ a. $5\frac{5}{8}$ b. 4 c. $3\frac{5}{16}$ d. $3\frac{41}{63}$

13. $5\frac{2}{3} + 7\frac{1}{2}$ a. $12\frac{7}{6}$ b. $12\frac{2}{3}$ c. $13\frac{1}{6}$ d. $13\frac{1}{3}$

14. $8\frac{3}{4} + 6\frac{1}{8}$

15. $9\frac{3}{5} + 4\frac{5}{8}$

16. $7\frac{4}{7} + 8\frac{5}{8}$ a. $15\frac{11}{16}$ b. $16\frac{11}{56}$ c. $15\frac{14}{56}$ d. $16\frac{45}{56}$

17. $10\frac{2}{5} + 9\frac{3}{5}$ a. $19\frac{5}{5}$ b. $19\frac{6}{5}$ c. 20 d. 21

18. $12\frac{1}{3} + 8\frac{4}{5}$

LESSON 25 – Adding Fractions and Mixed Numbers

Add one more fraction or mixed number to the group.

Example 1: $\dfrac{2}{3} + \dfrac{1}{2} + \dfrac{3}{4}$

LCM (2, 3, 4) = 12

$$\dfrac{2}{3} = \dfrac{8}{12}$$
$$\dfrac{1}{2} = \dfrac{6}{12}$$
$$\dfrac{3}{4} = \dfrac{9}{12}$$
$$\overline{}$$
$$\dfrac{23}{12} = 1\dfrac{11}{12}$$

Example 2: $3\dfrac{2}{5} + 4\dfrac{1}{2} + 5\dfrac{3}{4}$

LCM (2, 4, 5) = 20

$$3\dfrac{2}{5} = 3\dfrac{8}{20}$$
$$4\dfrac{1}{2} = 4\dfrac{10}{20}$$
$$5\dfrac{3}{4} = 5\dfrac{15}{20}$$
$$\overline{}$$
$$12\dfrac{33}{20} = 13\dfrac{13}{20}$$

Example 3: $4\dfrac{2}{3} + 5\dfrac{1}{2} + 2\dfrac{3}{5}$

LCM (2, 3, 5) = 30

$$4\dfrac{2}{3} = 4\dfrac{20}{30}$$
$$5\dfrac{1}{2} = 5\dfrac{15}{30}$$
$$2\dfrac{3}{5} = 2\dfrac{18}{30}$$
$$\overline{}$$
$$11\dfrac{53}{30} = 12\dfrac{23}{30}$$

Try these for extra practice. Remember to reduce your answer where necessary.

1. $5\frac{3}{5} + 6\frac{3}{4} + 6\frac{1}{2}$

2. $2\frac{1}{2} + 5\frac{2}{9} + 4\frac{2}{3}$

3. $\frac{4}{5} + \frac{7}{12} + \frac{1}{6}$ a. none of these   b. $2\frac{13}{15}$ c. $1\frac{11}{20}$ d. $\frac{43}{60}$

4. $\frac{3}{7} + \frac{21}{35} + \frac{1}{10}$ a. none of these   b. $1\frac{9}{70}$ c. $\frac{79}{140}$ d. $\frac{25}{52}$

5. $\frac{1}{5} + \frac{7}{20} + \frac{1}{8}$ a. $1\frac{7}{20}$ b. $\frac{3}{11}$ c. none of these   d. $\frac{27}{80}$

6. $\frac{1}{3} + \frac{7}{15} + \frac{1}{10}$ a. $\frac{9}{10}$ b. $\frac{9}{28}$ c. none of these   d. $1\frac{4}{5}$

7. $2\frac{1}{5} + 9\frac{3}{4} + 6\frac{1}{2}$

8. $5\frac{1}{2} + 8\frac{1}{9} + 3\frac{2}{3}$

9. $8\frac{1}{5} + 3\frac{2}{9} + 7\frac{2}{3}$

10. $7\frac{1}{3} + 7\frac{7}{16} + 9\frac{1}{4}$

11. $6\frac{1}{2} + 4\frac{1}{9} + 5\frac{2}{3}$

12. $\frac{2}{7} + \frac{19}{21} + \frac{1}{6}$

13. $\frac{3}{5} + \frac{13}{20} + \frac{1}{8}$

14. $4\frac{2}{3} + 8\frac{1}{4} + 3\frac{1}{2}$

15. $2\frac{6}{7} + 4\frac{4}{9} + 2\frac{1}{3}$

16. $6\frac{1}{2} + 5\frac{4}{9} + 5\frac{2}{3}$

17. $\frac{2}{3} + \frac{1}{15} + \frac{1}{10}$ a. $\frac{5}{6}$ b. $\frac{5}{12}$ c. $1\frac{2}{3}$ d. none of these

18. $\frac{4}{7} + \frac{11}{21} + \frac{1}{6}$ a. none of these   b. $2\frac{11}{21}$ c. $\frac{53}{84}$ d. $\frac{8}{17}$

19. $\frac{2}{7} + \frac{3}{14} + \frac{1}{4}$ a. none of these   b. $\frac{3}{4}$ c. $1\frac{1}{2}$ d. $\frac{21}{25}$

20. $6\frac{1}{7} + 2\frac{5}{9} + 4\frac{1}{3}$

21. $\frac{6}{7} + \frac{14}{35} + \frac{1}{10}$

22. $\frac{2}{3} + \frac{3}{6} + \frac{1}{4}$

23. $\frac{4}{5} + 1\frac{3}{8} + 5\frac{2}{5}$

24. $8\frac{2}{3} + 7\frac{1}{4} + 3\frac{5}{6}$

25. $5\frac{7}{9} + 3\frac{2}{27} + 8\frac{2}{9}$

26. $9\frac{1}{4} + 7\frac{2}{3} + 5\frac{3}{8}$

27. $1\frac{1}{5} + 2\frac{1}{4} + 3\frac{1}{3}$

28. $8\frac{5}{6} + 9\frac{7}{12} + 10\frac{3}{4}$

29. $4\frac{2}{5} + 6\frac{3}{16} + 8\frac{17}{20}$

30. $5\frac{7}{8} + 6\frac{3}{4} + 7\frac{13}{16}$

31. $7\frac{5}{7} + 4\frac{2}{5} + 11\frac{16}{35}$ a. none of these   b. $22\frac{1}{7}$ c. $23\frac{4}{7}$ d. $24\frac{1}{7}$

32. $3\frac{5}{8} + 4\frac{5}{12} + 6\frac{5}{6}$ a. none of these   b. $14\frac{17}{18}$ c. $14\frac{2}{3}$ d. $14\frac{7}{8}$

33. $1\frac{3}{5} + 2\frac{7}{10} + 6\frac{4}{5}$ a. none of these   b. $10\frac{1}{10}$ c. $11\frac{1}{5}$ d. $9\frac{9}{10}$

LESSON 26 – Subtracting Mixed Numbers

Example 1:
$$4\frac{2}{3} = 4\frac{4}{6}$$
$$- 3\frac{1}{2} = 3\frac{3}{6}$$
$$1\frac{1}{6}$$

Example 2:
$$15\frac{1}{2} = 15\frac{2}{4}$$
$$- 11\frac{1}{4} = 11\frac{1}{4}$$
$$4\frac{1}{4}$$

Example 3:
$$8\frac{1}{4} = 8\frac{3}{12} = 7\frac{15}{12} \longleftarrow$$
$$- 5\frac{2}{3} = 5\frac{8}{12} = 5\frac{8}{12}$$
$$2\frac{7}{12}$$

Cannot subtract 8 from 3. Need to borrow.

$$8\frac{3}{12} = 8 + \frac{3}{12} = 7 + 1 + \frac{3}{12} = 7 + \frac{12}{12} + \frac{3}{12} = 7 + \frac{15}{12} = 7\frac{15}{12}$$

The shortcut is to borrow 1 from the whole number. Add the numerator and denominator together and place the sum over the original denominator.

$$8\frac{3}{12} = 7\,\frac{3 + 12}{12} = 7\frac{15}{12}$$

Example 4:
$$12\frac{1}{5} = 12\frac{4}{20} = 11\frac{24}{20}$$
$$- 7\frac{1}{4} = 7\frac{5}{20} = 7\frac{5}{20}$$
$$4\frac{19}{20}$$

Example 5:
$$16\frac{3}{8} = 16\frac{15}{40} = 15\frac{55}{40}$$
$$- 7\frac{4}{5} = 7\frac{32}{40} = 7\frac{32}{40}$$
$$8\frac{23}{40}$$

Now try these.

1. $14\frac{1}{6} - 1\frac{7}{8}$ a. $12\frac{7}{24}$ b. 13 c. $\frac{11}{686}$ d. $12\frac{17}{48}$

2. $11 - 1\frac{2}{3}$ a. $9\frac{3}{7}$ b. $\frac{8}{199}$ c. $9\frac{1}{3}$ d. 10

3. $8\frac{1}{2} - 1\frac{2}{3}$ a. $6\frac{1}{2}$ b. $6\frac{5}{6}$ c. 7 d. $\frac{5}{77}$

4. $4\frac{1}{4} - 1\frac{8}{9}$ a. 3 b. $2\frac{17}{36}$ c. $\frac{1}{193}$ d. $2\frac{13}{36}$

5. $9\frac{1}{2} - 1\frac{4}{5}$ a. $7\frac{7}{10}$ b. $7\frac{4}{5}$ c. $\frac{6}{77}$ d. 8

6. $15\frac{1}{8} - 1\frac{8}{9}$ a. $13\frac{17}{72}$ b. 14 c. $\frac{12}{1097}$ d. $13\frac{7}{36}$

7. $7\frac{1}{3} - 1\frac{3}{4}$

8. $8\frac{1}{7} - 1\frac{7}{8}$

9. $15\frac{1}{9} - 1\frac{4}{5}$

10. $6\frac{1}{6} - 1\frac{3}{4}$

11. $9\frac{1}{8} - 1\frac{5}{6}$

12. $13\frac{1}{5} - 1\frac{8}{9}$

13. $5\frac{1}{5} - 3\frac{2}{3}$

14. $9\frac{1}{3} - 4\frac{3}{4}$

15. $23\frac{3}{10} - 15\frac{2}{3}$

16. $16\frac{3}{5} - 14\frac{7}{8}$

17. $16\frac{3}{5} - 12$

18. $33 - 18\frac{7}{9}$

19. $12\frac{3}{4} - 7\frac{7}{8}$

LESSON 27 – Adding and Subtracting Fractions

Some review of both adding and subtracting fractions.

1. Subtract: $2\frac{2}{5} - 1\frac{3}{7}$ a. $1\frac{1}{35}$ b. $\frac{34}{35}$ c. $1\frac{1}{7}$ d. $\frac{24}{35}$

2. Add: $3\frac{3}{5} + 1\frac{3}{7}$ a. $5\frac{1}{35}$ b. $4\frac{1}{2}$ c. 5 d. $7\frac{1}{3}$

3. $3\frac{3}{4} + 2\frac{11}{12} - 2\frac{1}{3}$

4. $4\frac{1}{6} + 4\frac{1}{12} - 3\frac{2}{3}$

5. Subtract: $2 - 1\frac{5}{9}$

6. Subtract: $4\frac{3}{8} - 2\frac{1}{3}$

7. Add: $6\frac{4}{5} + 8\frac{3}{7}$

8. Add: $8\frac{3}{4} + 5\frac{3}{10}$

9. Subtract: $6\frac{1}{5} - 3\frac{5}{6}$

10. Subtract: $7\frac{1}{7} - 1\frac{1}{4}$

11. $5\frac{3}{4} + 3\frac{1}{12} - 3\frac{1}{6}$

12. $2\frac{1}{6} + 4\frac{11}{12} - 2\frac{3}{4}$

13. Add: $3\frac{1}{2} + 2\frac{2}{3}$ a. $5\frac{3}{7}$ b. $5\frac{4}{7}$ c. $6\frac{1}{6}$ d. 6

14. Add: $7\frac{1}{4} + 1\frac{1}{8}$ a. $8\frac{1}{8}$ b. $8\frac{3}{8}$ c. $8\frac{3}{16}$ d. 9

15. Add: $3\frac{2}{5} + 1\frac{1}{9}$ a. $4\frac{23}{45}$ b. $7\frac{9}{14}$ c. 5 d. $4\frac{3}{14}$

16. Add: $7\frac{1}{9} + 1\frac{1}{2}$ a. $8\frac{3}{13}$ b. $6\frac{10}{13}$ c. 9 d. $8\frac{11}{18}$

17. Subtract: $4\frac{2}{3} - 3\frac{7}{8}$ a. $1\frac{11}{48}$ b. $\frac{19}{24}$ c. $\frac{7}{48}$ d. $1\frac{1}{16}$

18. Subtract: $2\frac{3}{4} - 1\frac{9}{10}$ a. $1\frac{3}{80}$ b. $\frac{49}{80}$ c. $1\frac{3}{16}$ d. $\frac{17}{20}$

19. Subtract: $3\frac{1}{4} - 1\frac{1}{3}$ a. $2\frac{1}{72}$ b. $1\frac{3}{4}$ c. $1\frac{11}{12}$ d. $2\frac{5}{72}$

20. $2\frac{1}{3} + 4\frac{1}{12} - 4\frac{1}{6}$

21. $3\frac{2}{5} + 5\frac{1}{3} - 6\frac{1}{10}$

22. Add: $17\frac{3}{4} + 15\frac{2}{3}$

23. $16\frac{1}{6} + 3\frac{7}{12} - 10\frac{3}{8}$

24. Subtract: $8\frac{7}{9} - 3\frac{8}{11}$

25. $14\frac{3}{7} + 21\frac{4}{7} - 15\frac{3}{8}$

26. Add: $12\frac{3}{4} + 9\frac{3}{8}$

27. $18\frac{2}{7} + 9\frac{5}{8} - 23\frac{23}{28}$

28. $7\frac{2}{3} + 8\frac{1}{2} + 5\frac{11}{12} + 6\frac{5}{6}$

29. Add: $8\frac{2}{3} + 7\frac{1}{2}$

30. $25 + 16\frac{3}{4} - 9\frac{1}{5}$

31. Subtract: $8\frac{3}{7} - 4\frac{5}{8}$

32. Subtract: $9\frac{1}{4} - 6\frac{5}{9}$

33. $12\frac{2}{5} + 6\frac{7}{8} - 3\frac{6}{7}$

34. $8\frac{3}{5} + 7\frac{2}{3} - 10\frac{4}{15}$

35. $17\frac{9}{13} + 11\frac{3}{26} - 18\frac{1}{2}$

LESSON 28 – More Adding and Subtracting of Mixed Numbers
Removal of Parentheses Before Solving

Example 1: $\left(6\frac{2}{3} + 4\frac{1}{5}\right) - 5\frac{1}{3}$

Work inside the parentheses first.

$$
\begin{array}{rcl}
6\frac{2}{3} & = & 6\frac{10}{15} \\
+\,4\frac{1}{5} & = & +\,4\frac{3}{15} \\
\hline
10\frac{13}{15} & = & 10\frac{13}{15} \\
-\,5\frac{1}{3} & = & -\,5\frac{5}{15} \\
\hline
& & 5\frac{8}{15}
\end{array}
$$

Example 2: $\left(5\frac{3}{8} + 4\frac{1}{4}\right) - 3\frac{11}{12}$

$$
\begin{array}{rclclcl}
5\frac{3}{8} & = & 5\frac{3}{8} \\
+\,4\frac{1}{4} & = & +\,4\frac{2}{8} \\
\hline
9\frac{5}{8} & = & 9\frac{15}{24} & = & 8\frac{39}{24} \\
-\,3\frac{11}{12} & = & -\,3\frac{22}{24} & = & -\,3\frac{22}{24} \\
\hline
& & & & 5\frac{17}{24}
\end{array}
$$

Example 3: $\left(5\frac{1}{6} - 2\frac{1}{2}\right) - 1\frac{3}{4}$

$$
\begin{array}{rclclcl}
5\frac{1}{6} & = & 5\frac{1}{6} & = & 4\frac{7}{6} \\
-\,2\frac{1}{2} & = & -\,2\frac{3}{6} & = & -\,2\frac{3}{6} \\
\hline
2\frac{4}{6} & = & 2\frac{8}{12} & = & 1\frac{20}{12} \\
-\,1\frac{3}{4} & = & -\,1\frac{9}{12} & = & -\,1\frac{9}{12} \\
\hline
& & & & \frac{11}{12}
\end{array}
$$

Example 4: $(7\frac{4}{9} - 3\frac{5}{8}) + 2\frac{2}{3}$

$$
\begin{array}{rclcl}
7\frac{4}{9} & = & 7\frac{32}{72} & = & 6\frac{104}{72} \\
-3\frac{5}{8} & = & -3\frac{45}{72} & = & -3\frac{45}{72} \\
\hline
& & 3\frac{59}{72} & = & 3\frac{59}{72} = \\
& & +2\frac{2}{3} & = & +2\frac{48}{72} = \\
\hline
& & 5\frac{107}{72} & = & 6\frac{35}{72}
\end{array}
$$

Try these:

1. $(3\frac{1}{5} + 5\frac{9}{20}) - 5\frac{3}{4}$

2. $(2\frac{1}{4} + 4\frac{1}{12}) - 4\frac{2}{3}$

3. $(4\frac{1}{4} - 2\frac{1}{12}) + 2\frac{5}{6}$

4. $(5\frac{1}{4} + 3\frac{1}{8}) - 3\frac{1}{2}$

5. $(4\frac{1}{5} - 3\frac{11}{30}) + 4\frac{5}{6}$

6. $(5\frac{2}{3} + 2\frac{1}{12}) - 5\frac{5}{6}$

7. $(3\frac{1}{4} + 5\frac{1}{20}) - 3\frac{3}{5}$

8. $(2\frac{1}{6} + 4\frac{7}{30}) + 2\frac{3}{5}$

9. $(4\frac{1}{3} - 2\frac{1}{12}) + 3\frac{3}{4}$

10. $(3\frac{1}{5} + 5\frac{1}{10}) - 4\frac{1}{2}$

11. $(4\frac{2}{3} + 4\frac{1}{15}) - 5\frac{4}{5}$

12. $(8\frac{2}{5} - 3\frac{1}{2}) - 4\frac{3}{4}$

13. $(4\frac{1}{4} + 4\frac{1}{12}) - 3\frac{2}{3}$

14. $(5\frac{2}{5} + 5\frac{3}{20}) - 2\frac{3}{4}$

15. $(3\frac{1}{4} + 3\frac{7}{20}) - 4\frac{4}{5}$

16. $(3\frac{2}{5} - 2\frac{1}{15}) + 3\frac{2}{3}$

17. $(4\frac{1}{6} - 3\frac{5}{12}) + 2\frac{3}{4}$

18. $(5\frac{1}{4} + 4\frac{1}{8}) - 4\frac{1}{2}$

19. $(5\frac{1}{8} - 2\frac{1}{6}) - 1\frac{5}{12}$

20. $(2\frac{1}{6} + 4\frac{17}{30}) - 4\frac{4}{5}$

21. $(3\frac{3}{4} + 4\frac{5}{8}) - 2\frac{7}{16}$

22. $(8\frac{5}{8} - 3\frac{7}{8}) + 5\frac{2}{3}$

23. $(4\frac{3}{5} - 2\frac{7}{10}) + 6\frac{4}{5}$

24. $(7\frac{2}{5} + 4\frac{1}{3}) - 6\frac{3}{8}$

25. $(5\frac{1}{3} - 3\frac{3}{5}) + 7\frac{1}{2}$

26. $(6\frac{2}{3} + 5\frac{4}{7}) - 2\frac{19}{21}$

27. $(4\frac{3}{7} + 5\frac{9}{14}) - 7\frac{4}{21}$

28. $(8\frac{4}{9} + 3\frac{1}{3}) - 7\frac{17}{18}$

LESSON 29 - Percents

Percent, by definition, is a ratio of a number to 100. The symbol for percent is: %.

Example 1: Express $\frac{24}{100}$ as a percent.

$$\frac{24}{100} = 24\%$$

Example 2: Express $\frac{3}{5}$ as a percent.

Step 1: Setup a proportion: $\frac{3}{5} = \frac{n}{100}$

Step 2: Solve the proportion.

$$\frac{3}{5} = \frac{n}{100}$$

$$5 \cdot n = 3 \cdot 100$$
$$5n = 300$$
$$\frac{5n}{5} = \frac{300}{5}$$
$$n = 60$$

Answer: 60%

Example 3: Express 20% as a fraction in lowest terms.

$$20\% = \frac{20}{100} = \frac{1 \cdot {}^{1}\cancel{20}}{5 \cdot {}_{1}\cancel{20}} = \frac{1}{5}$$

Example 4: Express 150% as a mixed number in lowest terms.

$$\frac{150}{100} = \frac{3 \cdot {}^{1}\cancel{50}}{2 \cdot {}_{1}\cancel{50}} = \frac{3}{2} = 1\tfrac{1}{2}$$

Example 5: If 65% of the peaches is ripe, what percent are not ripe?

$$100\% - 65\% = 35\%$$

Practice on percent.

Express as a percent.

1. $\frac{1}{4}$ 2. $\frac{5}{8}$ 3. $\frac{17}{25}$ 4. $\frac{53}{50}$

5. $2\frac{1}{4}$ 6. $5\frac{11}{22}$ 7. $\frac{3}{10}$ 8. $\frac{17}{20}$

9. $4\frac{1}{5}$ 10. $\frac{27}{8}$ 11. $\frac{12}{25}$ 12. $3\frac{3}{4}$

13. $\frac{103}{25}$ 14. $12\frac{1}{2}$ 15. $\frac{3}{5}$ 16. $\frac{3}{4}$

Express as a fraction in lowest terms or a mixed number in simplest form.

17. 72%   18. 165%   19. 14%   20. 40%

21. 175%   22. 19%   23. $5\frac{1}{2}$%   24. 450%

25. 60%   26. 38%   27. 94%   28. 132%

29. $83\frac{1}{3}$%   30. 220%   31. 55%   32. 86%

33.   An airplane has 65% of its seats filled on this flight. What percent are empty?

34.   If the team lost 5 of 20 games, what percent did they win?

35.   How may games did the football team win if they lost 20% of 15 games?

36.   If 200 people attended a conference and 60% were men, how many women attended the conference?

37.   If you received 90% on a twenty-question quiz, how many questions did you miss?

38.   25% of the 100 visitors at a museum are older than 62 years on a certain day. How many are younger than 62 years?

39.   A crowd of 5000 people attended a game. 2600 were cheering for the winning team. What percent is that?

40.   Chance of snowfall today is 40%. What is that in terms of a fraction in lowest terms?

LESSON 30 – Percent Computations

Example 1: 5 is what percent of 20?

$$\frac{5}{20} = \frac{n}{100}$$
$$20 \cdot n = 5 \cdot 100$$
$$20n = 500$$
$$\frac{20n}{20} = \frac{500}{20}$$
$$n = 25$$

Answer: 25%

Example 2: 8 is what percent of 40?

$$\frac{8}{40} = \frac{n}{100}$$
$$40 \cdot n = 8 \cdot 100$$
$$40n = 800$$
$$\frac{40n}{40} = \frac{800}{40}$$
$$n = 20$$

Answer: 20%

Example 3: What percent of 60 is 15?

$$\frac{15}{60} = \frac{n}{100}$$
$$60 \cdot n = 15 \cdot 100$$
$$60n = 1500$$
$$\frac{60n}{60} = \frac{1500}{60}$$
$$n = 25$$

Answer: 25%

Practice on percent computations.

1.   6 is what percent of 60?

           a. 0.1%   b. 10%   c. 6%   d. $\frac{1}{10}$%

2. 4 is what percent of 40?

3. 6 is what percent of 30?

4. 9 is what percent of 90?

5. 2 is what percent of 10?

6. 3 is what percent of 30?

7. 5 is what percent of 25?

   a. $\frac{1}{20}$%   b. 0.2%   c. 5%   d. 20%

8. 1 is what percent of 10?

   a. 1%   b. 0.1%   c. 10%   d. $\frac{1}{10}$%

9. 4 is what percent of 40?

   a. $\frac{1}{10}$%   b. 4%   c. 10%   d. 0.1%

10. 8 is what percent of 40?

    a. 20%   b. 0.2%   c. 8%   d. $\frac{1}{20}$%

11. 7 is what percent of 70?

    a. 10%   b. 7%   c. $\frac{1}{10}$%   d. 0.1%

12. 3 is what percent of 15?

13. What percent of 40 is 14?

14. What percent of 90 is 30?

15. What percent of 25 is 14?

16. What percent of 20 is 25?

17. What percent of 50 is 33?

18. What percent of 80 is 16?

19. 18 is what percent of 25?

20. What percent of 70 is 28?

LESSON 31 – Solving a Proportion

Example 1: $\dfrac{n}{4} = \dfrac{5}{10}$

Cross multiply, then divide by the coefficient of the variable.

$$\dfrac{n}{4} = \dfrac{5}{10}$$
$$10 \cdot n = 4 \cdot 5$$
$$10n = 20$$
$$\dfrac{10n}{10} = \dfrac{20}{10}$$

Answer: $n = 2$

Example 2: $\dfrac{n}{4} = \dfrac{5}{10}$

$$\dfrac{n}{4} = \dfrac{5}{10}$$
$$20 \cdot a = 10 \cdot 6$$
$$20a = 60$$
$$\dfrac{20a}{20} = \dfrac{60}{20}$$

Answer: $a = 3$

Example 3: $\dfrac{b}{5} = \dfrac{3}{10}$

$$10b = 15$$

Answer: $b = \frac{3}{2} = 1\frac{1}{2}$

Example 4: $\dfrac{a}{b} = \dfrac{c}{d}$

$$ad = bc \quad \leftarrow \quad \text{REMEMBER!!}$$

Example 5:   Yes or no: Is $\dfrac{5}{15} = \dfrac{6}{18}$ a proportion?

According to example 4, ad = c. Therefore, 5 • 18 must equal 15 • 6 in order for this to be classified as a proportion.

$$5 \cdot 18 \stackrel{?}{=} 15 \cdot 6$$
$$90 \stackrel{\checkmark}{=} 90$$

Yes, $\dfrac{5}{15} = \dfrac{6}{18}$ is a proportion.

Example 6:   Yes or no: Is $\frac{3}{7} = \frac{9}{20}$ a proportion?

$$3 \cdot 20 \overset{?}{=} 7 \cdot 9$$
$$60 \neq 63$$

Answer: $\frac{3}{7} = \frac{9}{20}$ is **not** a proportion.

Practice on proportions.

Solve the proportion.

1. $\frac{q}{2} = \frac{4}{16}$

2. $\frac{g}{5} = \frac{2}{50}$

3. $\frac{r}{4} = \frac{5}{80}$

4. $\frac{4}{5} = \frac{20}{x}$ a. x = 25   b. x = 10   c. x = 18   d. x = 34

5. $\frac{2}{6} = \frac{x}{18}$ a. x = 3   b. x = 6   c. x = 11   d. x = 5

6. $\frac{3}{x} = \frac{6}{14}$ a. x = 7   b. x = 4   c. x = 12   d. x = 10

7. $\frac{d}{5} = \frac{4}{100}$

8. $\frac{2}{4} = \frac{x}{12}$ a. x = 5   b. x = 6   c. x = 11   d. x = 3

9. $\frac{p}{4} = \frac{5}{80}$

10. $\frac{4}{5} = \frac{x}{15}$ a. x = 5   b. x = 21   c. x = 9   d. x = 12

11. How many of the following are proportions?

$$\frac{3}{4} = \frac{24}{32} \qquad \frac{3}{4} = \frac{18}{24} \qquad \frac{3}{4} = \frac{15}{32} \qquad \frac{3}{4} = \frac{21}{20}$$

12. How many of the following are proportions?

$$\frac{4}{7} = \frac{32}{56} \qquad \frac{4}{7} = \frac{40}{70} \qquad \frac{4}{7} = \frac{28}{49} \qquad \frac{4}{7} = \frac{36}{63}$$

13. $\frac{2}{x} = \frac{10}{35}$ a. x = 6   b. x = 10   c. x = 7   d. x = 12

14. $\frac{t}{5} = \frac{5}{125}$

15. $\frac{f}{4} = \frac{3}{32}$

16. $\frac{g}{2} = \frac{6}{24}$

17. $\frac{p}{3} = \frac{4}{24}$

18. $\frac{2}{5} = \frac{4}{x}$ a. x = 5   b. x = 14   c. x = 8   d. x = 10

19. How many of the following are proportions?

$$\frac{3}{5} = \frac{15}{25} \qquad \frac{3}{5} = \frac{21}{35} \qquad \frac{3}{5} = \frac{18}{30} \qquad \frac{3}{5} = \frac{12}{20}$$

20. How many of the following are proportions?

$$\frac{3}{5} = \frac{9}{15} \qquad \frac{3}{5} = \frac{6}{10} \qquad \frac{3}{5} = \frac{15}{25} \qquad \frac{3}{5} = \frac{12}{10}$$

21. Solve the proportion: $\frac{4}{x} = \frac{12}{21}$

a. x = 10   b. x = 6   c. x = 4   d. x = 7

22. $\frac{x}{5} = \frac{8}{10}$ a. x = 2   b. x = 4   c. x = 6   d. x = 8

23. $\frac{6}{12} = \frac{8}{x}$ a. x = 4   b. x = 8   c. x = 12   d. x = 16

24. $\frac{c}{d} = \frac{e}{f}$ a. cf = de   b. cd = ef   c. ce = df   d. $\frac{c}{d} = \frac{de}{f}$

LESSON 32 – Converting a Percent to a Decimal

To convert a percent to a decimal, locate the decimal point, move it **two places to the left** and remove the percent sign.

Example 1:   34%

> Step 1:   Locate the decimal point: 34.%
> Step 2:   Move the decimal two places to the left: .34.%
> Step 3:   Remove the percent sign: 0.34

Example 2:   123% = 123.% = 1.23

Example 3:   65% = 65.% = 0.65

Example 4: $73\frac{1}{2}$% = 73.5% = 0.735

Example 5:   450% = 450.% = 4.50 = 4.5   The zero is not needed.

To change a decimal to a percent, move the decimal point **two places to the right** and add the percent sign.

Example 6:   0.45

> Step 1: Move the decimal two places to the right: 0.45.
> Step 2: Add the percent sign: 45%

Example 7:   0.69 = 69%

Example 8:   1.34 = 134%

Example 9:   0.03 = 3%

Example 10:   0.138 = 13.8%

Example 11:   4.365 = 436.5%

Practice on converting percents to decimals. Write or convert each percents to a decimal numeral.

1. 82%   a. 0.82   b. 82   c. 8.2   d. 0.082

2. 8%

3. 94%

4. 9%

5. 34%   a. 34   b. 3.4   c. 0.34   d. 0.034

6. 46%   a. 0.046   b. 4.6   c. 0.46   d. 46

7. 5%

8. 6%

9. 41%

10. 86%

11. 3%

12. 52%   a. 0.052   b. 0.52   c. 5.2   d. 52

13. 35%

14. 18%

15. 55%

16. 64%

17. 71%   a. 0.071   b. 0.71   c. 71   d. 7.1

18. 89%   a. 0.089   b. 8.9   c. 0.89   d. 89

19. 135%

20. 750%

LESSON 33 – Decimals, Fractions, and Percents

Complete the table.

Fraction	Decimal	Percent
	0.15	
$\frac{7}{10}$		
		142%
$\frac{12}{25}$		
	0.4	

Solutions:

Fraction	Decimal	Percent
$\frac{15}{100} = \frac{3}{20}$	0.15	0.15 = 15%
$\frac{7}{10}$	0.7	0.70 = 70%
$1\frac{42}{100}$	142.% = 1.42	142%
$\frac{12}{25}$	48.% = 0.48	$\frac{12}{25} = \frac{n}{100}$ n = 48 48%
$\frac{40}{100} = \frac{2}{5}$	0.4	0.4 = 0.40 = 40%

Practice on decimals, fractions, and percents. Complete the chart. Reduce fractions to lowest terms where necessary.

	Fraction	Decimal	Percent
1.	_____	1.2	_____
2.	$\frac{7}{25}$	_____	_____
3.	_____	_____	230%
4.	$\frac{11}{20}$	_____	_____
5.	_____	_____	225%
6.	_____	0.45	_____
7.	$\frac{1}{10}$	0.1	_____

    a. 1%   b. none of these   c. 0.1%   d. 10%

	Fraction	Decimal	Percent
8.	_____	0.41	41%

    a. $\frac{41}{50}$ b. $\frac{41}{10}$ c. none of these   d. $\frac{41}{100}$

	Fraction	Decimal	Percent
9.	$\frac{18}{25}$	_____	72%

    a. 0.72   b. 7.2   c. 72   d. none of these

	Fraction	Decimal	Percent
10.	$\frac{17}{25}$	_____	68%

    a. 68   b. none of these   c. 0.68   d. 6.8

	Fraction	Decimal	Percent
11.	_____	_____	15%
12.	_____	1.7	_____
13.	$\frac{1}{5}$	_____	_____
14.	_____	_____	180%
15.	$\frac{4}{25}$	_____	16%

    a. 16   b. 1.6   c. 0.16   d. none of these

LESSON 34 – Fractional Part of a Number

Example 1: $3\frac{2}{3}$ of what number is 14?

$$3\frac{2}{3} \cdot N = 14 \text{ Convert the mixed number to an improper fraction.}$$

$$\frac{11}{3} \cdot N = 14$$

$$\frac{3}{11} \cdot \frac{11}{3} \cdot N = \frac{14}{1} \cdot \frac{3}{11} \quad \text{Multiply both sides by the reciprocal of } \frac{11}{3}.$$

Answer: $N = \frac{42}{11} = 3\frac{9}{11}$

Example 2: $1\frac{2}{5}$ of what number is $9\frac{1}{4}$?

$$1\frac{2}{5} \cdot N = 9\frac{1}{4}$$

$$\frac{7}{5} \cdot N = \frac{37}{4}$$

$$\frac{5}{7} \cdot \frac{7}{5} \cdot N = \frac{37}{4} \cdot \frac{5}{7}$$

Answer: $N = \frac{185}{28} = 6\frac{17}{28}$

Practice on fractional part of a number. Simplify answer where necessary.

1. $2\frac{5}{9}$ of what number is 15?

   a. $\frac{5}{69}$ b. $5\frac{20}{23}$ c. $\frac{23}{135}$ d. $13\frac{4}{5}$

2. $3\frac{2}{3}$ of what number is $1\frac{3}{4}$?

3. $1\frac{1}{6}$ of what number is 19?

4. $1\frac{2}{3}$ of what number is 14?

5. $1\frac{8}{15}$ of what number is 19?

6. $1\frac{2}{5}$ of what number is $1\frac{1}{10}$?

7. $2\frac{1}{2}$ of what number is $5\frac{2}{3}$?

73

8. $2\frac{1}{2}$ of what number is 16?

          a. $6\frac{2}{5}$ b. $1\frac{3}{5}$ c. $\frac{5}{32}$ d. $\frac{5}{8}$

9. $1\frac{7}{10}$ of what number is 14?

          a. $\frac{17}{140}$ b. $8\frac{4}{17}$ c. $\frac{7}{85}$ d. $12\frac{1}{7}$

10. $1\frac{6}{7}$ of what number is $1\frac{1}{4}$?

11. $1\frac{2}{3}$ of what number is 12?

          a. $7\frac{1}{5}$ b. $\frac{4}{5}$ c. $\frac{5}{36}$ d. $1\frac{1}{4}$

12. $1\frac{1}{2}$ of what number is 15?

          a. $\frac{2}{5}$ b. $2\frac{1}{2}$ c. $\frac{1}{10}$ d. 10

13. $1\frac{1}{10}$ of what number is 17?

          a. $15\frac{5}{11}$ b. $6\frac{8}{17}$ c. $\frac{17}{110}$ d. $\frac{11}{170}$

14. $2\frac{1}{3}$ of what number is $1\frac{2}{3}$?

15. $1\frac{3}{10}$ of what number is $5\frac{1}{2}$?

16. $5\frac{1}{2}$ of what number is $1\frac{2}{11}$?

17. $1\frac{1}{2}$ of what number is 13?

18. $1\frac{1}{10}$ of what number is 11?

19. $1\frac{6}{13}$ of what number is 19?

20. $2\frac{3}{5}$ of what number is 14?

21. $3\frac{1}{2}$ of what number is 12?

22. $1\frac{3}{4}$ of what number is 17?

23. $4\frac{2}{5}$ of what number is 40?

LESSON 35 – Decimal Part of a Number

Example 1:   What decimal part of 500 is 125?

$$\frac{125}{500} = \frac{{}^1\cancel{25} \cdot {}^1\cancel{5}}{{}_1\cancel{25} \cdot {}_4\cancel{20}} = \frac{1}{4} = 0.25$$

or
$$
\begin{array}{r}
0.25 \\
500 \overline{\smash{\big)}\ 125.00} \\
\underline{100\ 0} \\
25\ 00 \\
\underline{25\ 00} \\
0
\end{array}
$$

Example 2:   Two-fifths of what number is 90?

$$\frac{2}{5} \cdot N = 90$$

$$\frac{5}{2} \cdot \frac{2}{5} \cdot N = \frac{90}{1} \cdot \frac{5}{2}$$

Answer: $N = \dfrac{450}{2} = 225$

Example 3:   What decimal part of 200 is 85?

$$\frac{85}{200} = \frac{{}^1\cancel{5} \cdot 17}{{}_1\cancel{5} \cdot 40} = \frac{17}{40} = 0.425$$

or
$$
\begin{array}{r}
0.425 \\
200 \overline{\smash{\big)}\ 85.000} \\
\underline{80\ 0} \\
5\ 00 \\
\underline{4\ 00} \\
1\ 000 \\
\underline{1\ 000} \\
0
\end{array}
$$

Example 4:   Three-quarters of what number is 141?

$$\frac{3}{4} \cdot N = 141$$

$$\frac{4}{3} \cdot \frac{3}{4} \cdot N = \frac{141}{1} \cdot \frac{4}{3}$$

$$N = \frac{564}{3} = 188$$

Practice on decimal part of a number. Simplify answer where necessary.

1. What decimal part of 880 is 110?

2. What decimal part of 400 is 300?

3. Five-tenths of what number is 294?

4. Eight-tenths of what number is 96?

5. Seven-tenths of 150 is what number?
    a. 157   b. 105   c. 73.5   d. 103

6. Nine-tenths of 140 is what number?
    a. 126   b. 149   c. 124   d. 113.4

7. Three-tenths of 110 is what number?
    a. 9.9   b. 33   c. 113   d. 31

8. Six-tenths of what number is 192?

9. Four-tenths of what number is 186?

10. What decimal part of 1040 is 260?

11. What decimal part of 360 is 90?

12. Two-tenths of what number is 258?

13. Four-tenths of what number is 120?

14. Three-tenths of 180 is what number?
    a. 56   b. 54   c. 183   d. 16.2

15. Five-tenths of 50 is what number?
    a. 23   b. 55   c. 12.5   d. 25

16. What decimal part of 75 is 37.5?

17. Three-fifths of what number is 24?

18. Five-eighths of what number is 48?

19. What decimal part of 900 is 45?

20. Nine-tenths of what number is 72?

LESSON 36 – Percent as a Rate

Example 1:   Sixty percent of what number is 120?

Set up a percent proportion: $\dfrac{120}{N} = \dfrac{60}{100}$

Cross multiply: $60 \cdot N = 120 \cdot 100$

$$\dfrac{60}{60} \cdot N = \dfrac{120 \cdot 100}{60}$$

Answer: N = 200

$$\boxed{\begin{array}{c} \text{Percent Proportion} \\[4pt] \dfrac{\text{Part}}{\text{What Number}} = \% \end{array}}$$

Example 2:   45% of what number is 225?

$$\dfrac{225}{N} = \dfrac{45}{100}$$

$$45N = 22500$$

Answer: N = 500

Example 3:   What percent of 6 is 5?

$$\dfrac{5}{6} = \dfrac{N}{100}$$

$$6N = 500$$

Answer: $N = 83\frac{1}{3}\%$

Example 4:   160% of what number is 200?

$$\dfrac{200}{N} = \dfrac{160}{100}$$

$$160N = 20000$$

Answer: N = 125

Example 5:   125% of 80 is what?

$$\dfrac{N}{80} = \dfrac{125}{100}$$

$$100N = 10000$$

Answer: N = 100

Example 6:   24 is 60% of N

$$\dfrac{24}{N} = \dfrac{60}{100}$$

$$60N = 2400$$

$$N = 40$$

Answer: N = 40

Practice on percent as a rate.

1. Forty percent of what number is 114?

2. Twenty-five percent of what number is 67?

3. Two hundred fifty percent of 90 is what number?

4. What percent of 20 is 36?

5. Three hundred seventy percent of what number is 370?

6. One hundred seventy percent of 90 is what number?

7. What percent of 14 is 42?

8. One hundred ninety percent of what number is 570?

9. What percent of 5.6 is 14?
    a. 40%   b. 250%   c. 265%   d. 35%

10. What percent of 14.4 is 25.2?
    a. 190%   b. 52%   c. 57%   d. 175%

11. One hundred seventy percent of 70 is what number?

12. What percent of 4 is 9?

13. Two hundred ten percent of what number is 630?

14. Forty percent of what number is 84?

15. Twenty-five percent of what number is 46?

16. Twenty percent of what number is 36?

17. What percent of 2.4 is 4.2?
    a. 57%   b. 190%   c. 62%   d. 175%

18. One hundred twenty percent of 20 is what number?

19. What percent of 18 is 36?

20. Two hundred sixty percent of what number is 650?

LESSON 37 – More Percents

Example 1:   What percent of 300 is 45?

$$\frac{45}{300} = \frac{N}{100}$$
$$300N = 4500$$

Answer: N = 15%

Example 2:   150 is 30% of what number?

$$\frac{150}{N} = \frac{30}{100}$$
$$30N = 15000$$

Answer: N = 500

Example 3:   Find 225% of 72.

$$\frac{P}{72} = \frac{225}{100}$$
$$100P = 16200$$

Answer: P = 162

Example 4:   What number is 160% of 150?

$$\frac{N}{150} = \frac{160}{100}$$
$$100N = 24000$$

Answer: N = 240

Example 5:   Find 200% of 60.

$$\frac{N}{60} = \frac{200}{100}$$
$$100N = 12000$$

Answer: N = 120

Example 6:   144 is 60% of what number?

$$\frac{144}{N} = \frac{60}{100}$$
$$60N = 14400$$
$$N = 240$$

Answer: N = 240

Practice on more percents.

1. What percent of 1300 is 65?

2. What percent of 1800 is 36?

3. What percent of 1600 is 32?

4. 72 is 20% of what number?

5. 78 is 40% of what number?

6. 120 is 20% of what number?

7. 228 is 50% of what number?
   a. 456   b. 11.4   c. 114   d. 45.6

8. 114 is 60% of what number?
   a. 190   b. 68.4   c. 684   d. 1900

9. 180 is 20% of what number?
   a. 900   b. 36   c. 9000   d. 3.6

10. 150 is 40% of what number?

11. What percent of 1400 is 28?

12. 96 is 20% of what number?

13. 192 is 30% of what number?
    a. 57.6   b. 640   c. 64   d. 5.76

14. 144 is 40% of what number?
    a. 3600   b. 576   c. 360   d. 57.5

15. 288 is 20% of what number?

16. Find 170% of 60.

17. What number is 140% of 50?
    a. 70   b. 3571   c. 7000   d. 36

18. Find 140% of 50.

19. Find 190% of 30.

20. What number is 120% of 260?
          a. 217   b. 21,667   c. 31,200   d. 312

21. Find 140% of 20.

22. Find 170% of 80.

23. What number is 130% of 80?
          a. 104   b. 62   c. 10,400   d. 6154

24. What number is 140% of 190?
          a. 26,600   b. 136   c. 266   d. 13,571

25. Find 140% of 40.

26. Find 130% of 90.

27. What number is 140% of 270?
          a. 193   b. 378   c. 19,286   d. 37,800

28. What number is 130% of 220?
          a. 286   b. 169   c. 28,600   d. 16,923

29. Find 160% of 30.

30. Find 130% of 40.

31. What percent of 200 is 30?

32. 16 is 20% of what number?

33. Find 105% of 70.

34. What percent of 500 is 425?

35. Find 175% of 40.

36. 72 is 30% of what number?

37. What percent of 150 is 60?

38. Find 250% of 60.

39. Find 325% of 80.

40. 48 is 80% of what number?

LESSON 38 – Percentage Increases

Example 1:   What percent of 20 is 32?

$$\frac{32}{20} = \frac{N}{100}$$
$$20N = 3200$$

Answer: N = 160%

Example 2:   It costs $35 to fill up your car with gasoline.

    a. How much will it cost next year if gas increases by 20%?
    b. What is the dollar increase?

        a. $35 × 120% = $35 × 1.2 = $42
           or $35 + $35 • 20% =
                $35 + $35 • 0.2 =
                $35 + $7 = $42
        b. $42 – $35 = $7

Answer: a. $42
           b. $7

Example 3:   Fifteen percent of what number is 45?

$$\frac{45}{N} = \frac{15}{100}$$
$$15N = 4500$$

Answer: N = 300

Example 4:   A basket of tomatoes costs $7.50. The cost of the tomatoes increases by

        10%. What is the new price of the tomato basket?

    $7.50 × 110% = $7.50 × 1.10 = $8.25
      or $7.50 + $7.50 • 10% =
           $7.50 + $7.50 • 0.1 =
           $7.50 + $0.75 = $8.25

Answer: $8.25

Example 5:   Thirty percent of 500 is what?

$$\frac{N}{500} = \frac{30}{100}$$
$$100N = 15000$$
$$N = 150$$

Answer: N = 150

Practice on percentage increases.

1. What percent of 16 is 44?

2. What percent of 22 is 44?

3. Five percent of what number is 340?

4. Fifty percent of what number is 225?

5. Ten percent of what number is 200?

6. What percent of 2 is 4?

7. The cost of building a house increases 8% every year. If it costs $97,000 to build a house this year, what would it cost to build a house next year?

8. A department store purchases a dress for $95. To sell the dress to customers, the price is increased by 26%. What is the price of the dress?

9. Due to a sudden freeze, the cost of apples increased by 30% in one month. If the cost after the increase was 52¢ per pound, what was the cost before the increase?

10. A department store purchases a coat for $110. To sell the coat to customers, the price is increased by 26%. What is the price of the coat?

11. The cost of building a house increases 12% every year. If it costs $143,000 to build a house this year, what would it cost to build a house next year?

12. A department store purchases a coat for $95. To sell the coat to customers, the price is increased by 18%. What is the price of the coat?

13. The cost of building a house increases 20% every year. If it costs $123,000 to build a house this year, what would it cost to build a house next year?

14. Bob was earning $6.00 per hour. After working 12 months he received a raise of 20%. What was his hourly rate after the raise?

15. Due to a truck strike, the cost of bananas increased by 35% in one month. If the cost after the increase was 81¢ per pound, what was the cost before the increase?

LESSON 39 – Solving Proportions with Mixed Numbers

Example 1: $\dfrac{x}{1\frac{1}{2}} = \dfrac{\frac{3}{4}}{2\frac{2}{5}}$

$2\frac{2}{5}x = 1\frac{1}{2} \cdot \frac{3}{4}$ Cross multiply. $1\frac{1}{2} \cdot \frac{3}{4} = \frac{3}{2} \cdot \frac{3}{4} = \frac{9}{8}$

$\dfrac{12x}{5} = \dfrac{9}{8}$ Cross multiply again.

$96x = 45$ or $\dfrac{5}{12} \cdot \dfrac{12}{5} x = \dfrac{9}{8} \cdot \dfrac{5}{12}$

$x = \dfrac{45}{96} = \dfrac{15}{32}$ $x = \dfrac{\overset{3}{9}}{8} \cdot \dfrac{5}{\underset{4}{12}} = \dfrac{3}{8} \cdot \dfrac{5}{4} = \dfrac{15}{32}$

Answer: $\dfrac{15}{32}$

Example 2: $\dfrac{1\frac{3}{4}}{4\frac{1}{2}} = \dfrac{x}{7\frac{1}{3}}$

$4\frac{1}{2}x = 1\frac{3}{4} \cdot 7\frac{1}{3}$ $1\frac{3}{4} \cdot 7\frac{1}{3} = \dfrac{7}{\underset{2}{4}} \cdot \dfrac{\overset{11}{22}}{3} = \dfrac{77}{6}$

$\dfrac{9x}{2} = \dfrac{77}{6}$

$54x = 154$ or $\dfrac{2}{9} \cdot \dfrac{9}{2} x = \dfrac{77}{6} \cdot \dfrac{2}{9}$

$x = \dfrac{154}{54} = 2\frac{46}{54} = 2\frac{23}{27}$ $x = f(77,_{3}6) \cdot \dfrac{\overset{1}{2}}{9} = \dfrac{77}{27} = 2\frac{23}{27}$

Answer: $2\frac{23}{27}$

Practice on solving proportions with mixed numbers. Simplify answers where necessary.

1. $\dfrac{x}{3\frac{1}{5}} = \dfrac{\frac{1}{4}}{1\frac{1}{2}}$
                                  2. $\dfrac{\frac{1}{5}}{3\frac{1}{2}} = \dfrac{x}{2\frac{6}{7}}$

3.. $\dfrac{1\frac{2}{3}}{x} = \dfrac{1\frac{5}{7}}{\frac{3}{5}}$
          a. $\frac{12}{7}$ b. $\frac{100}{21}$ c. $\frac{7}{12}$ d. $\frac{21}{100}$

4. $\dfrac{4\frac{2}{7}}{1\frac{1}{5}} = \dfrac{x}{\frac{2}{3}}$ 　　　a. $\dfrac{50}{21}$ 　　b. $\dfrac{54}{7}$ 　　c. $\dfrac{7}{54}$ 　　d. $\dfrac{24}{7}$

5. $\dfrac{\frac{5}{4}}{x} = \dfrac{1\frac{3}{5}}{1\frac{3}{7}}$ 　　　a. $\dfrac{125}{112}$ 　　b. $\dfrac{20}{7}$ 　　c. $\dfrac{7}{20}$ 　　d. $\dfrac{64}{35}$

6. $\dfrac{1\frac{1}{5}}{2\frac{2}{3}} = \dfrac{x}{\frac{2}{3}}$ 　　　a. $\dfrac{5}{24}$ 　　b. $\dfrac{3}{10}$ 　　c. $\dfrac{24}{5}$ 　　d. $\dfrac{32}{15}$

7. $\dfrac{x}{3\frac{3}{4}} = \dfrac{\frac{2}{3}}{4\frac{2}{3}}$ 　　　　　　　　8. $\dfrac{x}{\frac{3}{4}} = \dfrac{3\frac{3}{7}}{2\frac{1}{4}}$

9. $\dfrac{\frac{1}{3}}{2\frac{1}{2}} = \dfrac{x}{3\frac{3}{5}}$ 　　　a. 3 　　b. $\dfrac{12}{25}$ 　　c. 27 　　d. $\dfrac{1}{27}$

10. $\dfrac{4\frac{4}{7}}{4\frac{1}{2}} = \dfrac{x}{\frac{1}{2}}$ 　　　　　　　11. $\dfrac{2\frac{1}{2}}{3\frac{3}{4}} = \dfrac{4\frac{2}{5}}{x}$

12. $\dfrac{5\frac{7}{8}}{3\frac{2}{3}} = \dfrac{x}{4\frac{1}{5}}$ 　　　　　　　13. $\dfrac{x}{4\frac{4}{5}} = \dfrac{6\frac{2}{3}}{5\frac{1}{8}}$

14. $\dfrac{6\frac{3}{5}}{3\frac{1}{8}} = \dfrac{3\frac{3}{5}}{x}$ 　　　　　　　15. $\dfrac{2\frac{5}{9}}{x} = \dfrac{4\frac{4}{5}}{7\frac{1}{6}}$

LESSON 40 – Easy Story Problems

Example 1:   Joe spent $6.18 for 6 pens. How much did each one cost?

$$
\begin{array}{r}
1.03 \\
6\,\overline{\smash)6.18} \\
\underline{6} \\
18 \\
\underline{18} \\
0
\end{array}
$$

Answer: $1.03 each

Example 2:   CDs are on sale: 2 for $9. You bought 7 CDs. How much change did you

get from two $20 bills?

$$\frac{7}{2} \times \$9 = \frac{\$63}{2} = \$31.50 \text{ for 7 CDs}$$

$2(\$20) = \$40$

$\$40 - \$31.50 = \$8.50$ change

Answer: $8.50 change

Example 3:   Sixteen DVDs cost $132.50. How much did each DVD cost?

$$
\begin{array}{r}
8.30 \\
16\,\overline{\smash)132.80} \\
\underline{128} \\
4\,8 \\
\underline{4\,8} \\
0
\end{array}
$$

Answer: $8.30 each

Practice problems on easy story problems.

1.   A discount audio store advertised a collection of hits from the 1990's on 11 compact discs for $47.52. What is the cost of each disc?

2.   A discount audio store advertised a collection of classical music on 14 compact discs for $85.40. What is the cost of each disc?

3. Amber spent $4.27 at the stationery store. If she bought 7 erasers, how much did each eraser cost?

    a. $0.63   b. $0.61   c. $0.60   d. $29.89

4. Adrian spent $2.55 at the stationery. If he bought 5 note pads, how much did each note pad cost?

    a. $0.50   b. $0.53   c. $12.75   d. $0.51

5. A discount audio store advertised a collection of heavy metal music on 24 compact discs for $105.36. What is the cost of each disc?

6. A discount audio store advertised a collection of hits from the 1970's on 13 compact discs for $67.86. What is the cost of each disc?

7. Adrian spent $4.08 at the stationery. If he bought 8 pencils, how much did each pencil cost?

    a. $0.51   b. $12.75   c. $0.50   d. $0.53

8. A discount audio store advertised a collection of hits from the 1960's on 14 compact discs for $79.24. What is the cost of each disc?

9. Adrianna spent $1.24 at the stationery. If she bought 4 pens, how much did each pen cost?

    a. $0.29   b. $0.31   c. $4.96   d. $0.30

10. Patricia paid $595 for 7 nights at a hotel. What was the nightly rate for her room?
    a. $43 per night   b. $170 per night
    c. $85 per night   d. $4165 per night

11. The Smith's paid $292.50 for 26 weeks of the Morning News. How much was that each week?

12. If a dozen cans of beans cost $6.48, how much does one can of beans cost?

13. On your first math exam you scored a 90%. What is the minimum score on the next exam to maintain an 85% average?

14. Each group (trio) enters a building and two exit the back. How many people remain in the building after four groups enter?

15. A pound (16 ounces) of cheese cost $5.12. How much does a quarter pound cost?

16. Six cans of juice cost $0.66. Two dozen cans would then cost how much?

17. How much does it cost to outfit 12 players on a team if it costs $42 per player?

18. Ten people form a group in a parade. After one block, three exit and five enter the group. This continues for four more blocks. How many are in the group after five blocks?

19. Sam bought a $12.99 DVD and 6 compact discs at 75¢ each. How much change did he receive from a $20 bill?

20. At the flea market, Mary sold seven items which totaled $35.25. The purchaser gave her two twenty dollar bills. How much change did Mary give the purchaser?

21. How much change did Sandi receive after she purchased three CD's at $3.99 each and gave the clerk a $5 and $10 bill?

22. Rumple lost his wallet which had a $10 bill, three $5 bills and seventeen $1 bills. How much did Rumple lose?

23. Four bars of chocolate cost 80¢ each. How many more chocolate bars can Amy purchase if she only has a $5 bill?

24. Shasta purchased three cases of bottled water for $3.50 each. After she gave the clerk a $20 bill, she remembered to get a twelve-pack of Mountain Dew. That item cost $3.75. How much change did she receive?

25. Benny bought six pairs of flip-flops for $47.94 total. What change did he receive from three $20 bills? Could he buy any additional flip-flops with his change?

26. Candy bought three jars of ketchup for $1.20 each along with two dispensers of mustard for $1.50 each. How much did the condiments costs?

ANSWERS

LESSON 1
1. b 2. d 3. d 4. c
5. hundred-thousand
6. ten-million 7. ten-billion
8. ten-thousand 9. a 10. d

LESSON 2
1. c 2. a 3. b 4. 45,170
5. 58,300 6. 46,000 7. b 8. a
9. a 10. 57,000 11. 47,300
12. 60,000 13. 4730 14. 9300
15. d 16. c 17. b 18. 68,000
19. 44,500 20. 6,000,000
21. 74,200 22. c 23. b 24. c
25. 64,400 26. 86,000
27. 147,000 28. 8,000,000
29. 3460 30. 15,700
31. 43,880,000 32. 23,500
33. 175,000

LESSON 3
1. d 2. b 3. 15 4. 32 5. 7592
6. 5142 7. 2970 8. d 9. c
10. 3667 11. 6232 12. 5484
13. 35 14. d 15. b 16. 2356
17. 7929 18. 5813 19. 58
20. 110,579 21. 75 22. $50.05

LESSON 4
1. 78,687 2. 46,257 3. d 4. b
5. b 6. 66,742 7. 1435
8. 382 × 6 = 2292 9. 40,915
10. a 11. b 12. b 13. 73 R 19
14. 85 R 36 15. 8 16. 12
17. 29 18. 9 19. 43 R 54
20. 25 R 55 21. 10
22. 1048 R 3

23. $6 \overline{)54}$, 54 ÷ 7, $\frac{54}{6}$ 24. a

25. 15

LESSON 5
1. a, $33.25 + $13.73 =
$$\begin{array}{r} \$ 33.25 \\ \text{or} + 13.73 \\ \hline \$ 46.98 \end{array}$$
$46.98, or
2. b, $41.12 + $12.52 =
$$\begin{array}{r} \$ 41.12 \\ \text{or} + 12.52 \\ \hline \$ 53.64 \end{array}$$
$53.64, or
3. 52, 79 − 27 = 52, or
$$\begin{array}{r} 79 \\ - 27 \\ \hline 52 \end{array}$$
4. $0.82, $10 − $9.18 = $0.82,
$$\begin{array}{r} \$ 10.00 \\ \text{or} \quad - 9.18 \\ \hline \$ 0.82 \end{array}$$

5. $3.58,
$$\begin{array}{r} \$ 5.00 \\ - 1.42 \\ \hline \$ 3.58 \end{array}$$

6. 6484 candy bars,
$$\begin{array}{r} 3119 \\ + 3365 \\ \hline 6484 \end{array}$$

7. 17,000 seats,
$$\begin{array}{r} 24,000 \\ - 7,000 \\ \hline 17,000 \end{array}$$

8. a,
$$\begin{array}{r} 1433 \\ + 5994 \\ \hline 7427 \end{array}$$

9. 12,000 seats,
$$\begin{array}{r} 17,000 \\ - 5,000 \\ \hline 12,000 \end{array}$$

10. c,
$$\begin{array}{r} \$ 64.00 \\ + 12.50 \\ \hline \$ 76.50 \end{array}$$
11. $2.01 12. 4
13. $8 14. first

LESSON 6
1. d 2. b 3. 8.77 4. 32.08
5. 28.35 6. b 7. c 8. 15.38
9. 8.645 10. 0.0076
11. 0.0336 12. d 13. 0.09
14. 0.084 15. 0.0516
16. 0.0792

17. 4.10,
$$\begin{array}{r} 2.29 \\ + 0.61 \\ \hline 2.90 \end{array} \quad \begin{array}{r} 7.00 \\ - 2.90 \\ \hline 4.10 \end{array}$$
18. 726.94 19. 49.692
20. 58.53 21. 2.37

LESSON 7
1. 0.21 2. 0.7 3. 0.06 4. 0.41
5. 0.71 6. a 7. d 8. 0.42
9. 0.71 10. 0.21 11. 0.41
12. 0.32 13. b 14. d 15. c
16. a 17. 0.002 18. 0.06
19. 2.3 20. 0.092 21. 0.375
22. 2.54 23. 2.664 24. 5.06
25. 1.25 26. 1251.2 27. 73.26
28. b 29. 0.64 30. 28.31
31. 142.56 32. 63.72 33. 7500
34. 0.002 35. 56.1432

LESSON 8
1. c 2. b 3. $\frac{83}{100}$ 4. $\frac{27}{100}$ 5. 0.50

6. $\frac{11}{100}$ 7. $\frac{47}{100}$ 8. d 9. b 10. a

11. d 12. c 13. d 14. b 15. $\frac{79}{100}$

16. $\frac{29}{100}$ 17. a 18. d 19. c 20. b

21. d 22. a 23. c 24. $\frac{73}{100}$

25. $\frac{37}{100}$ 26. 0.70 27. $43\frac{13}{20}$

28. 16.4 29. $123\frac{9}{20}$

LESSON 9
1. no 2. a 3. c 4. c 5. b 6. d
7. a 8. no 9. yes 10. no 11. no
12. yes 13. d 14. b 15. a 16. c
17. no 18. no 19. no 20. yes
21. yes 22. no 23. yes 24. yes
25. no 26. yes 27. yes 28. yes
29. no 30. yes

LESSON 10
1. 31 2. 83 3. c 4. b 5. 73
6. 47 7. b 8. a
9. 18, 21, 26, 10
10. 12, 16, 18, 6 11. d 12. d
13. b 14. 22, 26, 32, 12 15. a
16. 4, 6, 8, 9, 10, 12, 14, 15, 16, 18
17. 20, 21, 22, 24, 25, 26, 27, 28, 30
18. 23, 29, 31, 37, 41, 43, 47
19. 8, 35, 77, 54, 33, 57
20. 60, 62, 63, 64, 65, 66, 68, 69, 70
21. 4; 13, 17, 19, 23 22. false
23. 3; 4, 6, 8
24. 1, 2, 3, 4, 5, 6, 7, 8, 9
25. 4, 6 26. composite

LESSON 11
1. a 2. d 3. 2•2•5•5•7
4. 3• 5•5•7 5. 2•2•3•3•5•7
6. c 7. b 8. a 9. 2•2•2•3•5•5•7
10. 2•2•3•5•5•11
11. 2•2•3•5• 7 12. d 13. b
14. b 15. 2•2• 2•5•11
16. 2•2•2•3•3•5•5• 11
17. 2•2•3•3•5•5•7 18. b 19. d
20. b 21. c 22. b 23. d 24. d
25. 2^2 • 3 • 143 26. 2^2 • 5 • 19
27. 2^2 • 5 • 73 28. 2^3 • 5^3
29. 3 • 7 • 37 30. 2^5 • 3 • 5 •11

LESSON 12
1. a 2. c 3. a 4. a 5. a 6. b 7. c
8. 10, 20, 30, 40, 50, 60
9. 12, 24, 36, 48, 60
10. 11, 22, 33, 44, 55, 66, 77, 88, 99

LESSON 13
1. b 2. a 3. 8 4. 12 5. 30 6. 24
7. 72 8. 10 9. 18 10. c 11. 24
12. 12 13. 18 14. c 15. 12
16. 10 17. c 18. 60 19. 30
20. 45 21. 24 22. 120 23. 24
24. 50 25. 72 26. 10 27. 56
28. 20 29. 90 30. 60

LESSON 14
1. b 2. b 3. 8 4. 25 5. 16 6. 64
7. 125 8. 49 9. d 10. a 11. c
12. 36 13. 64 14. 343 15. 100
16. 81 17. 256 18. 64 19. 512
20. 729 21. 1000 22. 625
23. 121 24. 225 25. 144
26. 169 27. $4 \cdot x \cdot x$
28. $16 \cdot a \cdot a \cdot a$
29. $(4 \cdot x \cdot 4 \cdot x)$ or $(4 \cdot 4 \cdot x \cdot x)$
30. $20 \cdot b \cdot b \cdot b$ 31. $9cc$
32. $10 \cdot a \cdot a \cdot b \cdot b$
33. $4 \cdot x \cdot x \cdot y \cdot z$
34. $16 \cdot x \cdot x \cdot x \cdot x$

LESSON 15
1. 8 2. 5 3. 9 4. 16 5. 6 6. 16
7. 12 8. 16 9. 38 10. 42 11. 22
12. 12 13. 6 14. 6 15. 2 16. 24
17. 24 18. 12 19. 6 20. 30

LESSON 16
1. a 2. c 3. c 4. b 5. c 6. b 7. b
8. a 9. c 10. b 11. c 12. $\frac{3}{5}$ 13. $\frac{5}{6}$
14. $\frac{2}{5}$ 15. $\frac{3}{4}$ 16. b 17. c 18. d
19. b 20. $\frac{13}{20}$ 21. $\frac{1}{3}$ 22. $\frac{2}{5}$ 23. $\frac{2}{3}$
24. $\frac{7}{8}$ 25. $\frac{3}{10}$ 26. $\frac{2}{3}$ 27. $\frac{3}{8}$ 28. $\frac{5}{12}$

LESSON 17
1. a. $-\frac{1}{9}$, b. 9

2. Zero does not have a reciprocal because division by zero is undefined.

3. b 4. c 5. b 6. a. 8, b. $-\frac{1}{8}$

7. a. -2, b. $\frac{1}{2}$ 8. d 9. c 10. -5
11. b 12. b 13. d 14. b 15. c
16. c 17. d 18. a 19. a 20. d

LESSON 18
1. $\frac{8}{35}$ 2. $\frac{4}{63}$ 3. d 4. d 5. $\frac{1}{5}$ 6. $\frac{1}{3}$

7. $\frac{4}{55}$ 8. $\frac{4}{15}$ 9. $\frac{2}{3}$ 10. $\frac{24}{77}$ 11. $1\frac{1}{3}$

12. a 13. $\frac{1}{5}$ 14. $\frac{4}{21}$ 15. $\frac{8}{55}$ 16. 4

17. a 18. a 19. $2\frac{1}{4}$ 20. b 21. $2\frac{1}{2}$

22. $3\frac{2}{3}$ 23. $5\frac{1}{2}$ 24. $1\frac{1}{2}$ 25. $\frac{1}{3}$

26. $1\frac{1}{6}$ 27. 4 28. 1 29. $\frac{3}{4}$ 30. 6

LESSON 19
1. b 2. d 3. b 4. b 5. b 6. b 7. c
8. a 9. a 10. a 11. d 12. b
13. c 14. b 15. d

LESSON 20
1. d 2. c 3. $\frac{14}{5}$ 4. $\frac{3}{2}$ 5. a 6. $\frac{5}{2}$

7. $\frac{11}{4}$ 8. $\frac{17}{5}$ 9. a 10. b 11. d

12. $\frac{21}{5}$ 13. $\frac{51}{7}$ 14. $\frac{35}{4}$ 15. $\frac{35}{8}$

16. $\frac{20}{3}$ 17. $\frac{29}{5}$ 18. $\frac{42}{11}$ 19. $\frac{47}{8}$

LESSON 21
1. d 2. $\frac{16}{27}$ 3. $\frac{1}{2}$ 4. $\frac{10}{21}$ 5. $\frac{4}{15}$ 6. $\frac{1}{3}$

7. $\frac{28}{45}$ 8. $\frac{27}{55}$ 9. $\frac{10}{33}$ 10. d 11. c

12. $\frac{1}{3}$ 13. $\frac{16}{45}$ 14. $\frac{1}{4}$ 15. $\frac{16}{45}$ 16. $\frac{20}{33}$

17. $\frac{8}{27}$ 18. $\frac{8}{35}$ 19. $\frac{4}{27}$ 20. a

21. $\frac{40}{77}$ 22. $\frac{8}{15}$

LESSON 22
1. $\frac{2}{15}$ 2. $3\frac{1}{3}$ 3. $\frac{2}{3}$ 4. 50 5. $2\frac{1}{3}$

6. $1\frac{1}{8}$ 7. $\frac{6}{7}$ 8. $\frac{3}{25}$ 9. $3\frac{8}{9}$ 10. $1\frac{7}{9}$

11. $\frac{5}{12}$ 12. $1\frac{17}{64}$ 13. 1 14. $\frac{20}{63}$

15. $\frac{5}{6}$ 16. $1\frac{1}{5}$ 17. 21 18. $1\frac{3}{35}$

19. $2\frac{4}{13}$ 20. $1\frac{2}{15}$

LESSON 23
1. a 2. a 3. c 4. d 5. $\frac{25}{33}$ 6. $2\frac{1}{13}$

7. $\frac{28}{45}$ 8. $15\frac{1}{2}$ 9. $\frac{1}{48}$ 10. $\frac{3}{10}$

11. $37\frac{5}{7}$ 12. $49\frac{30}{49}$ 13. $31\frac{13}{27}$

14. $42\frac{9}{10}$ 15. $\frac{3}{4}$ 16. $\frac{44}{133}$ 17. b

18. c 19. $18\frac{9}{32}$ 20. 6

LESSON 24
1. $13\frac{4}{15}$ 2. $12\frac{19}{30}$ 3. $12\frac{5}{12}$ 4. $3\frac{17}{24}$

5. $15\frac{14}{15}$ 6. c 7. c 8. a 9. $15\frac{5}{12}$

10. $7\frac{4}{15}$ 11. c 12. d 13. c

14. $14\frac{7}{8}$ 15. $14\frac{9}{40}$ 16. b 17. c

18. $21\frac{2}{15}$

LESSON 25
1. $18\frac{17}{20}$ 2. $12\frac{7}{18}$ 3. c 4. b 5. c

6. a 7. $18\frac{9}{20}$ 8. $17\frac{5}{18}$ 9. $19\frac{4}{45}$

10. $24\frac{1}{48}$ 11. $16\frac{5}{18}$ 12. $1\frac{5}{14}$

13. $1\frac{3}{8}$ 14. $16\frac{5}{12}$ 15. $7\frac{40}{63}$

16. $17\frac{11}{18}$ 17. a 18. a 19. b

20. $13\frac{2}{63}$ 21. $1\frac{5}{14}$ 22. $1\frac{5}{12}$

23. $7\frac{23}{40}$ 24. $19\frac{3}{4}$ 25. $17\frac{2}{27}$

26. $22\frac{7}{24}$ 27. $6\frac{47}{60}$ 28. $29\frac{1}{6}$

29. $19\frac{7}{16}$ 30. $20\frac{7}{16}$ 31. c 32. d

33. a

LESSON 26
1. a 2. c 3. b 4. d 5. a 6. a
7. $5\frac{7}{12}$ 8. $6\frac{15}{56}$ 9. $13\frac{14}{45}$ 10. $4\frac{5}{12}$

11. $7\frac{7}{24}$ 12. $11\frac{14}{45}$ 13. $1\frac{8}{15}$

14. $4\frac{7}{12}$ 15. $7\frac{19}{30}$ 16. $1\frac{29}{40}$ 17. $4\frac{3}{5}$

18. $14\frac{2}{9}$ 19. $4\frac{7}{8}$

LESSON 27
1. b 2. a 3. $4\frac{1}{3}$ 4. $4\frac{7}{12}$ 5. $\frac{4}{9}$ 6.
$2\frac{1}{24}$ 7. $15\frac{8}{35}$ 8. $14\frac{1}{20}$ 9. $2\frac{11}{30}$ 10.
$5\frac{25}{28}$ 11. $5\frac{2}{3}$ 12. $4\frac{1}{3}$ 13. c 14. b
15. a 16. d 17. b 18. d 19. c
20. $2\frac{1}{4}$ 21. $2\frac{19}{30}$ 22. $33\frac{5}{12}$ 23. $9\frac{3}{8}$
24. $5\frac{5}{99}$ 25. $20\frac{5}{8}$ 26. $22\frac{1}{8}$ 27.
$4\frac{5}{56}$ 28. $28\frac{11}{12}$ 29. $16\frac{1}{6}$ 30. $32\frac{11}{20}$
31. $3\frac{45}{56}$ 32. $2\frac{25}{36}$ 33. $15\frac{117}{280}$ 34. 6
35. $10\frac{4}{13}$

LESSON 28
1. $2\frac{9}{10}$ 2. $1\frac{2}{3}$ 3. 5 4. $4\frac{7}{8}$ 5. $5\frac{2}{3}$ 6.
$1\frac{11}{12}$ 7. $4\frac{7}{10}$ 8. 9 9. 6 10. $3\frac{4}{5}$ 11.
$2\frac{14}{15}$ 12. $\frac{3}{20}$ 13. $4\frac{2}{3}$ 14. $7\frac{4}{5}$ 15. $1\frac{4}{5}$
16. 5 17. $3\frac{1}{2}$ 18. $4\frac{7}{8}$ 19. $1\frac{13}{24}$ 20.
$1\frac{14}{15}$ 21. $5\frac{15}{16}$ 22. $10\frac{5}{12}$ 23. $8\frac{7}{10}$
24. $5\frac{43}{120}$ 25. $9\frac{7}{30}$ 26. $9\frac{1}{3}$ 27. $2\frac{37}{42}$
28. $3\frac{5}{6}$

LESSON 29
1. 25% 2. 62.5% 3. 68%
4. 106% 5. 225% 6. 550%
7. 30% 8. 85% 9. 420%
10. 337.5% 11. 48%
12. 375% 13. 412%
14. 1250% 15. 60% 16. 75%
17. $\frac{18}{25}$ 18. $1\frac{13}{20}$ 19. $\frac{7}{50}$ 20. $\frac{2}{5}$
21. $1\frac{3}{4}$ 22. $\frac{19}{100}$ 23. $\frac{11}{200}$ 24. $4\frac{1}{2}$
25. $\frac{3}{5}$ 26. $\frac{19}{50}$ 27. $\frac{47}{50}$ 28. $1\frac{8}{25}$
29. $\frac{5}{6}$ 30. $2\frac{1}{5}$ 31. $\frac{11}{20}$ 32. $\frac{43}{50}$
33. 35% empty 34. 75%
35. 12 games 36. 80 women
37. 2 38. 75 39. 52% 40. $\frac{2}{5}$

LESSON 30
1. b 2. 10% 3. 20% 4. 10% 5.
20% 6. 10% 7. d 8. c 9. c 10.
a 11. a 12. 20% 13. 35% 14.
$33\frac{1}{3}$% 15. 56% 16. 125% 17.
66% 18. 20% 19. 72% 20.
40%

LESSON 31
1. $\frac{1}{2}$ 2. $\frac{1}{5}$ 3. $\frac{1}{4}$ 4. a 5. b 6. a 7. $\frac{1}{5}$
8. b 9. $\frac{1}{4}$ 10. d 11. 2 12. 4 13. c
14. $\frac{1}{5}$ 15. $\frac{3}{8}$ 16. $\frac{1}{2}$ 17. $\frac{1}{2}$ 18. d
19. 4 20. 3 21. d 22. b 23. d
24. a

LESSON 32
1. a 2. 0.08 3. 0.94 4. 0.09
5. c 6. c 7. 0.05 8. 0.06
9. 0.41 10. 0.86 11. 0.03
12. b 13. 0.35 14. 0.18

15. 0.55 16. 0.64 17. b 18. c
19. 1.35 20. 7.5

LESSON 33
1. $\frac{6}{5} = 1\frac{1}{5}$, 120% 2. 0.28, 28%
3. $\frac{23}{10} = 2\frac{3}{10}$, 2.3 4. 0.55,55%
5. $\frac{9}{4} = 2\frac{1}{4}$, 2.25 6. $\frac{9}{20}$,45% 7. d
8. d 9. a 10. c 11. $\frac{3}{20}$, 0.15
12. $\frac{17}{10} = 1\frac{7}{10}$, 170%
13. 0.2, 20% 14. $\frac{9}{5} = 1\frac{4}{5}$, 1.8
15. c

LESSON 34
1. b 2. $\frac{21}{44}$ 3. $16\frac{2}{7}$ 4. $8\frac{2}{5}$ 5. $12\frac{9}{23}$
6. $\frac{11}{14}$ 7. $2\frac{4}{15}$ 8. a 9. b 10. $\frac{35}{52}$
11. a 12. d 13. a 14. $\frac{5}{7}$ 15. $4\frac{3}{13}$
16. $\frac{26}{121}$ 17. $8\frac{2}{3}$ 18. 10 19. 13
20. $5\frac{5}{13}$ 21. $3\frac{3}{7}$ 22. $9\frac{5}{7}$ 23. $9\frac{1}{11}$

LESSON 35
1. $\frac{1}{8} = 0.125$ 2. $\frac{3}{4} = 0.75$ 3. 588
4. 120 5. b 6. a 7. b 8. 320
9. 465 10. $\frac{1}{4} = 0.25$
11. $\frac{1}{4} = 0.25$ 12. 1290 13. 300
14. b 15. d 16. 0.5 17. 40
18. 76.8 19. 0.05 20. 80

LESSON 36
1. 285 2. 268 3. 225 4. 180%
5. 100 6. 153 7. 300% 8. 300
9. b 10. d 11. 119 12. 225%
13. 300 14. 210 15. 184
16. 180 17. d 18. 24 19. 200%
20. 250

LESSON 37
1. 5% 2. 2% 3. 2% 4. 360
5. 195 6. 600 7. a 8. a 9. a
10. 375 11. 2% 12. 480 13. b
14. c 15. 1440 16. 102 17. a
18. 70 19. 57 20. d 21. 28
22. 136 23. a 24. c 25. 56
26. 117 27. b 28. a 29. 48
30. 52 31. 15% 32. 80
33. 73.5 34. 85% 35. 70
36. 240 37. 40% 38. 150
39. 260 40. 60

LESSON 38
1. 275% 2. 200% 3. 6800
4. 450 5. 2000 6. 200%
7. $104,760 8. $119.70 9. 40¢
per pound 10. $138.60
11. $160,160 12. $112.10
13. $147,600 14. $7.20
15. 60¢ per pound

LESSON 39
1. $\frac{8}{15}$ 2. $\frac{8}{49}$ 3. c 4. a 5. a 6. b
7. $\frac{15}{28}$ 8. $\frac{8}{7}$ 9. b 10. $\frac{32}{63}$ 11. $6\frac{3}{5}$
12. $6\frac{321}{440}$ 13. $6\frac{10}{41}$ 14. $1\frac{31}{44}$
15. $3\frac{1057}{1296}$

LESSON 40
1. $4.32 2. $6.10 3. b 4. d
5. $4.39 6. $5.22 7. a 8. $5.66
9. b 10. c 11. $11.25
12. $0.54 13. 80%
14. 4 people 15. $1.28
16. $15.84 17. $504
18. 20 people 19. $2.51
20. $4.75 21. $3.03 22. $42
23. 2 24. $5.75
25. $12.06, yes 26. $6.60

INDEX